P9-CSE-203

NATURE'S NETHER REGIONS

ALSO BY MENNO SCHILTHUIZEN

Frogs, Flies, and Dandelions: The Making of Species

The Loom of Life: Unravelling Ecosystems

NATURE'S NETHER REGIONS

What the Sex Lives
of Bugs, Birds, and Beasts
Tell Us About
Evolution, Biodiversity,
and Ourselves

....

Menno Schilthuizen

VIKING

VIKING
Published by the Penguin Group
Penguin Group (USA) LLC
375 Hudson Street
New York, New York 10014

USA | Canada | UK | Ireland | Australia | New Zealand | India | South Africa | China
penguin.com
A Penguin Random House Company
First published by Viking Penguin, a member of Penguin Group (USA) LLC, 2014
Copyright © 2014 by Menno Schilthuizen

Penguin supports copyright. Copyright fuels creativity, encourages diverse voices, promotes free speech, and creates a vibrant culture. Thank you for buying an authorized edition of this book and for complying with copyright laws by not reproducing, scanning, or distributing any part of it in any form without permission. You are supporting writers and allowing Penguin to continue to publish books for every reader.

Illustrations by Jaap Vermeulen. Copyright © 2014 by Menno Schilthuizen.

LIBRARY OF CONGRESS CATALOGING-IN-PUBLICATION DATA
Schilthuizen, Menno.
Nature's nether regions : what the sex lives of bugs, birds, and beasts tell us about evolution, biodiversity, and ourselves / Menno Schilthuizen.
pages cm
Includes bibliographical references and index.
ISBN 978-0-670-78591-9
1. Generative organs—Evolution. 2. Sexual behavior in animals—Evolution. I. Title.
QL876.S35 2014
573.6—dc23
2013047833

Printed in the United States of America

1 3 5 7 9 10 8 6 4 2

Set in Warnock Pro
Designed by Francesca Belanger

Contents

NATURE'S NETHER REGIONS

. . . yes and his heart was going like mad and yes I said yes I will Yes.

—James Joyce, *Ulysses*

Preliminaries

Not too long ago, the Netherlands Natural History Museum was housed in a lofty, cavernous building in the historical center of Leiden. Generations of biology students took their zoology classes there, in its two-tier lecture theater over the monumental staircase. During the less captivating parts on crustacean leg structure or mollusk shell dentition, their gaze would have wandered off to the two features that made this lecture room unforgettable. First, its abundant display of antlers of deer, antelope, and other hoofed animals, hundreds of them, suspended from the walls. Second, the huge painting from 1606 of a beached sperm whale that hung over the lectern. On an otherwise nondescript Dutch beach lies the Leviathan, its beak agape, its limp tongue touching the sand. A smattering of well-dressed seventeenth-century Dutchmen stand around the beast. Prominently located, and closest to the dead whale, stand a gentleman and his lady. With a lewd smile, face turned toward his companion, the gentleman points at the two-meter-long penis of the whale that sticks out obscenely from the corpse. Centuries of smoke-tanned varnish cannot conceal the look of bewilderment in her eyes.

These few square feet of canvas, strategically placed in the painting's golden ratio, exemplify two things. First, the unassailable fact (supported by millennia of bathroom graffiti, centuries of suggestive postcards, and decades of Internet images) that humans find genitals endlessly fascinating. Their own, but by extension those of other creatures, too. The amazing diversity in shape, size, and function of the reproductive organs of animals has been an eternal source of wonder, making bestsellers of the 1953 book *The Sex Life of Wild Animals,* the 1980s classroom wall poster *Penises of the Animal Kingdom* (over twenty thousand copies sold), and the Sundance Channel series *Green Porno*—short films starring a sanguine Isabella Rossellini enacting the copulation of various animals.

The second point that may be underscored with this seventeenth-century sperm whale penis is the curious observation that the public fascination with genitalia was, until very recently at least, not matched by equally intensive scientific inquiry. The lofty offices down the corridor from this lecture theater housed scores of biologists quietly cataloging the world's biodiversity. In good classificatory tradition, they would painstakingly draw, measure, photograph, and describe the minutiae of the genitals and distinguishing features of the reproductive organs of any new insect, spider, or millipede they would discover—and yet never stop to wonder how these private parts evolved.

We really have Darwin to blame for this. In his next-greatest book, *The Descent of Man, and Selection in Relation to Sex* (1871), Darwin explains how secondary sexual characteristics—like colorful bird plumage, the prongs on beetles' heads, and the antlers of deer—have been shaped not by natural selection (adaptation to the environment) but by sexual selection: adaptation to the preferences of the other sex. He denies the primary sexual characteristics entry to his theory by categorically stating that sexual selection is not concerned with the genitalia or primary sexual organs—which, after all, are merely functional, not fanciful. So the diversity of all those antlers on the walls of the museum lecture room had been a tradition of evolutionary biology since Darwin, but investigating the evolution of the business end of things—of which the centerpiece of that seventeenth-century painting is just one prominent example—hadn't.

It took until 1979 for evolutionary biology to start paying attention to genitalia. In that year, Jonathan Waage, an entomologist from Brown University, published a short paper in *Science* on the damselfly penis. He demonstrated that this minuscule penis carried a miniature spoon that, during mating, cleaned out the female's vagina, scooping out any remaining sperm from previous males. It was an eye-opener as well as a sperm-scooper. For the first time, here was proof that animal genitals are not just mundane sperm-depositing and sperm-receiving organs, but are sites where a sexual selection of sorts goes on. After all, during damselfly evolution, males with the best sperm-scoopers had left more descendants.

The time was ripe for this paper. When I interviewed Waage about those early days, he recalled how, in the years leading up to his sperm-scooper discovery, he had been influenced by the quiet revolution that biology faculties worldwide were undergoing at the time—a sea change brought about by George C. Williams's book *Adaptation and Natural Selection* and by Richard Dawkins's popularization of it, *The Selfish Gene.* People began to do away with the false notion that evolution works "for the good of the species" (an outdated concept, echoes of which can be heard even today in nature documentaries). Instead, they began viewing evolution correctly, as the effect of a kind of reproductive selfishness, in which it is all about the success of an individual in carrying its genes into the next generation. Evolution does not "care" about the species. And if a sperm-scooper would scupper the chances of competing males, then that is what evolution would favor. Waage was one of the first scientists to start asking the questions that mattered for how evolution works. And since evolution is all about reproduction, no wonder Waage and other modern biologists would sooner or later find themselves closely inspecting genitals.

In that same revolutionary era, other young biologists began asking similar questions. One of them was a certain undergraduate biology student who in the 1960s was earning some extra cash with menial tasks in the depot of Harvard University's Museum of Comparative Zoology. His job was to top up the alcohol in jars with pickled animals and to organize unsorted spider specimens. Picking up spider identification guides, the student began wondering why spider species are so often distinguished by the way their genitals are formed. Asking around in the museum, he was told that that is just the way it is. The genitals of different species of animals, be they spiders, spittlebugs, or Spanish fly, are often widely different, even if the species are one another's close relatives and look identical on the outside. Probably, his seniors told him, the genetic differences also accidentally affect the shape of the genitals. Very useful if you want to identify spiders, but probably quite meaningless biologically. The student, unconvinced but not in a position to argue, shelved the question in the back of his mind, graduated, and went on to become a productive and successful tropical biologist at the Smithsonian Institution's Tropical Research Institute in Panama.

That student's name was Bill Eberhard. And when, many years later, the issue of *Science* with Waage's damselfly penis article landed on his desk, that old conundrum from his undergraduate days let out a little muffled cry from beneath many layers of mental clutter. Perhaps genitalia, in spiders as well as other animals, differ so much because each is a different kind of sperm-scooper? As it happened, Eberhard was about to begin a six-month stint as a visiting scientist at the University of Michigan, which gave him the opportunity to spend some weeks in the library.

There, he pulled off one of those rare feats of biological unification. It is often not realized that the basic source of inspiration in biology, namely the endless diversity of life, is also one of its greatest handicaps. Biologists, much more than, say, chemists or mathematicians, tend to be divided by invisible barriers. Those barriers are held in place by expertise with a particular kind of organism. More often than not, biologists identify themselves as entomologists if they work with insects, or as botanists if plants are their thing. Or even as copepodologists, coleopterologists, or cecidomyiidologists (if their creed be copepods, beetles, or cecidomyiid gnats, respectively). And each organism-based field has its own congresses, professional societies, and journals, further affirming separatism. Contrary to, for example, physicists, to whom a neutron is a neutron is a neutron, biologists are always unsure whether what applies to one kind of organism also applies to another—or, worse, they don't care about broad applicability at all. As ecologist Stephen Hubbell lamented, if Galileo had been a biologist, he would have spent his whole life documenting the trajectories of different animals thrown off the Leaning Tower of Pisa without ever coming up with gravitational acceleration.

Biology really moves ahead when somebody dares to cut across all those different subfields and look for general patterns. And that is precisely what Eberhard did when he cloistered himself in the University of Michigan library and began pulling books off the shelves on the genitalia of mice and moles, snails and snakes, weevils and whales. Four years later, in 1985, what had started as a little hobby project had turned into the 256-page Harvard University Press classic *Sexual Selection and Animal Genitalia.* In it, besides dazzling his reader with a sheer endless parade of wondrously shaped animal willies, Eberhard made two points. First, that genitals are bafflingly complex systems, far too complicated for the relatively simple task of depositing and receiving a droplet of sex cells. The male chicken flea, for example, has a "penis" that is actually a profusion of plates, combs, springs, and levers and looks more like an exploded grandfather clock than a syringe—whereas the latter should suffice if the organ's role were just to squirt sperm into the female. And the second point he made was that no body part in the animal kingdom evolves as fast as genitalia.

In his book, Eberhard argued that the reproductive organs of animals are under constant, intensive, and multitarget sexual selection—including the kind revealed by Waage, but certainly not limited to that. This is why they are so complex. This is also why they differ so much from species to species—a phenomenon that taxonomists (that special breed of biologist whose task it is to circumscribe, describe, name, and classify biodiversity) had been happily using throughout the twentieth century as an easy way to differentiate species. Animals' nether regions are the stages where an evolutionary play is performed that would have made even Darwin blush. An evolutionary play that had been totally ignored by generations of biologists—even though genitalia are probably the best body parts to illustrate the power of evolution.

And yet the evidence had been, quite literally, staring us in the face. Humans and our fellow primates do not shy from Eberhard's accelerated genital evolution. Forget forebrains, canine teeth, and opposable big toes: the largest anatomical differences between us and our closest relative, the chimpanzee, are found in our nethers. The human vagina is flanked by two pairs of skin folds, the labia minora and the labia majora. The clitoris is a two-winged structure lying along the walls of the vagina, and only the relatively small glans is visible externally, covered by the clitoral foreskin and lying at the point where the labia minora join. The chimpanzee vagina, on the other hand, lacks labia minora, has a larger and downward-pointing glans of the clitoris, and contains specialized tissue that makes the labia and the foreskin of the clitoris swell dramatically during the fertile phase of the menstrual cycle, causing the vagina to bulge out and increasing its operating depth by 50 percent. And on the other side of the sexual divide the differences between these two sister species are no less striking. The human penis is thick and blunt ended, boneless, has a ridge around the smooth glans, and has a foreskin. It has two corpora cavernosa, the sponge-like tissue that swells during erection. The chimp penis, by contrast, is thin and sharply pointed, carries a penis bone (baculum) inside, has no glans, no foreskin, and only one corpus cavernosum. Oh—and it carries lots of tiny tough spines along the sides.

In other words, the exaggerated diversity—biodiversity—in genital shapes that Eberhard had highlighted carries right up to our own species. The evidence for this pattern throughout the animal world is available in large quantities of respectable nineteenth- and twentieth-century tomes on comparative anatomy and systematic zoology, and yet before Eberhard nobody had bothered to explain it.

But this is not a book about Bill Eberhard. Rather, it is about the band of disciples that followed in his footsteps. Hundreds of scientists all over the world, myself included, have been inspired by Eberhard's book. Together, with our lab experiments, fieldwork, and computer simulations on a wide variety of organisms from primates to pack rats and from sea slugs to sexton beetles, we have nursed to life a brand-new discipline of evolutionary biology: a science of the genitals, if you will. And, as disciples and disciplines are wont, we have increasingly come to dispute the exact workings of genital evolution. Are penises internal courtship devices, as Eberhard would have it? Or are they used to combat rival males on the female's turf, as Waage showed? Or are male and female genitalia perhaps at loggerheads over who is in charge of fertilization, as people like the English zoologist Tracey Chapman think?

Despite these bones of contention, two things unite these scientists. First, a genuine desire to understand. To reconstruct the tortuous routes by which evolution has graced the animal kingdom with such a bewildering diversity of reproductive organs. And second, that same innate interest in all things sexual that is the reason why you are reading this book and also the reason why I wrote it.

Such fascination with private parts notwithstanding, by devoting an entire book to the field, and by not shying from the more complicated bits, I hope to rise above the giggly press genital researchers have been getting. I am not saying this book will be any less naughty in tone. Still, rather than being a vaudeville of juicy anecdotes fished from the nooks and crannies of animal weirdness, evolution of genitalia has, over the past twenty-five years, matured into a solid science where extreme biodiversity, advanced evolutionary theory, and elegant experimentation come together. My aim is to paint a portrait of this new branch of biology.

From time immemorial, we have taken the mechanics of sexual intercourse for granted. But the nitty-gritty of our own reproduction is anything but default. The evolution of our genitals has steered the evolution of our copulation behavior and vice versa, blessing (or saddling) us with just one of the possible outcomes of countless scenarios of complex evolutionary interactions, involving everything along the continuum between graceful dances and vicious arms races. Realizing this may make us better appreciate humans' place in the reproductive diversity of life.

Chapter 1

Define Your Terms!

This book is not about sex.

A puzzling statement, perhaps. You could have sworn that the preceding pages were strewn with words and phrases that in everyday parlance would be flagged as decidedly sex related. But then the meaning of biological terms in everyday parlance often is quite divorced from the same terms used by actual biologists. To a biologist—at least during working hours—"sex" does not mean the events leading up to and including the insertion of genitalia into somebody else's genitalia and/or additional orifices. Instead, it means something like "the exchange of DNA between two individuals." And exchanging DNA can be done in a multitude of ways, many of which do not involve any activity that the man or woman on the street would consider "sex."

Take bacteria, for instance. They regularly pick up bits of DNA from other bacteria, which they transfer to their own genetic machinery via finger-like protrusions called pili. They even take up and incorporate into their chromosomes whatever loose strands of DNA take their fancy as they encounter them in their microscopic environment. Such "bacterial sex," as microbiologists call it, is a far cry from the results one gets when typing "sex" into an Internet search engine. For starters, bacteria use sex—that is, looting the environment for bits of DNA code—not for procreation but to improve their own lot (so do many people on those Internet pages, but that's another matter). The DNA that bacteria mop up from their environment might contain genes that they can put to good use. To fix gaps in their own DNA, for example, or to feed on foods that their original DNA did not contain the digestive tools for. They don't do it to reproduce. For that, bacteria simply divide themselves—in the bacterial world, sex and reproduction are two entirely unrelated activities.

For most larger organisms, like ourselves, sex *is* a usual component of reproduction. We carry double sets of all our genes—one set inherited from Mom, one from Dad—produce eggs and sperm that contain single sets of those genes, and combine sperm with eggs to produce children with reconstituted double sets of genes. But there are many different ways in which organisms can make sure that eggs and sperm meet, and copulation is just one of them. Fixed on their reefs as they are, corals, for example, cannot consort with one another and thus are left with no option but to release their eggs and sperm into the waters and hope for the best—that is, fortuitous chance encounters between them. And the birch trees that line the streets of many a northern country pump billions of pollen into the air each spring, of which a very small fraction is wind-carried to the stamens of female catkins. Only a few hay fever sufferers realize that they are sneezing themselves through clouds of birch ejaculate.

Okay, you might say, so sex is perhaps a bit unconventional in such obscure things like microorganisms and corals, but surely most of the more familiar animals "have sex" to mix their DNA with that of their partner and produce babies, right? Sorry, no. Not necessarily. Pseudoscorpions, for example, don't. In these animals, which look like miniature scorpions but without the sting, males simply leave tiny stalked sperm-filled balloons scattered throughout their neighborhood. Females encounter these surprises and, if they feel so inclined, position their genital openings over them, squat down a little, and absorb them. And many species of springtail and salamander perform similarly impersonal sex. In fact, biologists think that this is the original system, and that genitalia evolved later to make the transfer of such sperm packages more efficient. What we, from our myopic human-centered perspective, consider "sex" is just one of the many ways that organisms have evolved to combine packaged DNA from one individual with that of another.

Another general misconception is that, in nature at least, sex and reproduction are synonymous. But they are not. We have just seen that bacteria have sex (that is, they mix foreign DNA with their own) but don't necessarily reproduce in the process. Conversely, there are lots of organisms that reproduce without sex. Bacteria, but also many plants, some parasitic wasps, stick and other insects, some lizards, as well as tiny aquatic creatures called bdelloid rotifers, to name but a few, almost always eschew sex. They consist entirely of females that simply give birth to cloned daughters that are genetically identical copies of themselves. No males, no exchange of DNA via sperm and eggs, and certainly no hanky-panky.

In fact, now that we are on the subject, biologists are still puzzled as to why there should be sex at all. Cloning yourself, as the animals in the previous paragraph do, is four times as efficient as sexual reproduction. First, you don't need to share your genes with those of a male (a twofold advantage); second, all your children can have babies, rather than only the female half (another twofold advantage). The fact that sex is so pervasive in nature means that there must be an enormous benefit to having sex over cloning yourself. And, no, in biology "the joy of sex" does not qualify as an advantage. Instead, you may be surprised to learn that biologists think that sexual reproduction evolved either as a way of outsmarting parasites or as a way to purge your DNA of harmful mutations.

The parasite theory goes as follows. Let's, for the sake of argument, imagine that humans were a clonal species. That Eve, so to speak, had never lain with her male companion but instead begot genetically identical daughters who then gave birth to clonally reproduced granddaughters and so on, until the whole world was populated by identical copies of Eve.

Enter a killer parasite. In a sexual species, such a deadly parasite—a virus, for example—would normally not be able to spread very far, because soon it would encounter individuals that were genetically so different from its first victims that it would need to mutate to overcome their immune systems. But in a clonal species, everybody is genetically identical, has the exact same weak spots, and is thus equally susceptible to the new parasite, which would spread like wildfire and kill off all clonal Eves in no time.

The benefits of clonal reproduction could thus be lost in one disastrous sweep of parasitic infection. A sexually reproducing animal or plant, on the other hand, does not run this risk, because all its offspring are genetically different (being random recombinations of the genes of both parents), so that even if a particularly mean parasite strikes, there would always be some offspring that are more resistant than others.

So there you have it: to stay one step ahead of fast-evolving parasites, the members of a species have to use sex to keep reshuffling their genes all the time. Since this is akin to getting nowhere fast, the parasite hypothesis is also known as the "Red Queen" hypothesis, after the Red Queen in Lewis

Carroll's *Through the Looking-Glass,* who tells Alice, "Now, *here,* you see, it takes all the running *you* can do, to keep in the same place."

Attractive as the parasite hypothesis may be, there is another popular (well, popular among evolutionary biologists) explanation for the benefits of sex: that it is a clever way to get rid of accumulated errors in your DNA. Each time DNA is copied—to produce a sperm or egg cell, for example, or during cloning—there is a small chance that one or a few letters in the DNA code will be misread by the copying machinery (which is only chemical, after all) and misincorporated into the copy. Occasionally, this leads to a T where first there was an A, or a C is accidentally replaced by a G, or perhaps an A is unintentionally doubled to AA or skipped altogether.

These "spelling errors" are sometimes harmless or even beneficial, but more often they will be flies in the genetic ointment. In a celibate organism that reproduces solely by copying itself, there is no way to prevent such harmful mutations from accumulating from one generation to the next, like making photocopies of photocopies of photocopies, which eventually leads to illegible text. Each daughter inherits the exact genome of her mother, flaws and all, and adds new ones of her own. Over many generations, lots of those little errors will have piled up in her descendants, and overall their genetic health will deteriorate.

Now, sex can prevent this. Of course, during the production of eggs and sperm such errors are also made and are inherited by sexually created offspring. But since the genetic shuffling during the production of eggs and sperm is a chance process, as is the combination of sperm and eggs to produce new organisms, some offspring will inherit lots of errors, and some none at all. This means that if the ones with fewer inherited DNA mistakes are slightly "fitter," they will be the ones surviving, thus purging each litter of the worst genetic flaws.

Scientists are still debating which of these two theories is more likely to explain the benefits of sex. What is beyond doubt, though, is that such benefits must exist. Without them, all creation would simply be cloning itself, and there would be no genders, no sperm, no eggs, no mating, no genitals, and certainly no popular science books about them. So it is important to keep in mind that sex, familiar and unavoidable as it may seem to us, is not the default way of reproducing in nature. It is Reproduction 2.0, a surprisingly complicated way that has evolved to avoid the encumbrances of straightforward cloning.

Whence She and He?

And there are more aspects about sex that seem to go without saying but, upon closer scrutiny, beg for an explanation. Take males and females. What is that all about? Nothing in the menu for sexual reproduction specifies that for the mixing of DNA two *different* kinds of individuals are required. Think about it: if there were only one gender and everybody could mate with everybody else, then finding a mate would be twice as easy while still keeping the genetic benefits of sex. What could possibly be the point of imposing a rule that says there must be two genders and you are allowed to reproduce only if you mix your genes with those of the other gender?

Nature is not bureaucratic, so there must be a good reason for such a strange decree. Not surprisingly, biologists disagree over what that reason may have been back in the deep recesses of the prehistory of life. They have come up with several theories, but the one with the best cards argues that separate sexes evolved to prevent war between organelles. I realize you must be frowning now. War? Organelles? Let me elaborate.

All organisms beyond the complexity of bacteria carry so-called organelles in their cells. These are tiny contraptions that perform important functions. An example is the green chloroplasts that sit in plant cells and that house the chlorophyll and the rest of the photosynthesis machinery. Although they seem to be purpose-built micromachines, such organelles are actually the stripped-down descendants of free-living bacteria that, at some time in the distant evolutionary past, invaded the cells of other organisms and began a joint venture with them. They still retain some independence: they have their own DNA and divide themselves.

And in this organelle independence lies the problem. During sex, one sex cell of one organism fuses with one sex cell of another organism. If both contribute their organelles to the daughter cell that is produced by this fusion, it will be populated by two types of organelles: one type from one parent and another, with probably slightly different organellar DNA, from the other parent. Since both types of organelles play the same role in the cell, evolution will favor those types that are best at competing against the intracellular rivals. This may mean that organelles would evolve to draw a lot of resources from their host cell to be able to divide more quickly than the organelles that they share the cell with, or even produce toxic substances to kill their rivals.

Having the inside of its cells turned into an organelle battlefield cannot be good for the host, so if sexual reproduction started off with the fusion of identical sex cells, sooner or later evolution came up with an improved system. In that system, some organisms made very small sex cells, which carried zero or very few organelles, and others made much larger sex cells with lots of organelles. When two small sex cells fused, they would not have enough organelles to start life. When two big sex cells fused, their organelles would engage in a war of attrition over cell domination. But when a small and a large sex cell fused, the organelles from the big partner would immediately swamp the few contributed by the small partner, and the rest of the life of the new organism would not be plagued by any more organelle warfare.

The evolutionary result of this cellular peace process was a system for sexual reproduction with two different kinds of organisms: one kind ("male") always producing small sex cells ("sperm") that contribute DNA to the offspring but no organelles, and another kind ("female") delivering large sex cells ("eggs") with DNA plus lots of organelles. It's a sobering thought that the whole system of separate males and females and the ensuing war of the sexes may have come about as a necessary complication to prevent an even more disastrous microscopic war inside our cells. In fact, as the closing chapter of this book will show, males and females do not even need to be in separate bodies. Hermaphrodite animals are male and female at the same time, equipped with masculine as well as feminine machineries, fertilize each other and yet, despite this equality, live even more bizarre sex lives than "regular" animals.

What Is Primary Anyway?

Ask any medical doctor what primary and secondary sexual characteristics are, and she will roll down a wall chart with a man and a woman in full frontal nudity and deftly point out the geography of both on the human body. Penis and scrotum with testicles in the man, and vagina in the woman are the primary sexual characteristics—at least as far as is visible without the aid of a scalpel or a speculum. Secondary sexual characteristics are lots of additional differences between men and women, scattered all over the body and ranging from breasts, hips, and stature to hair-loss patterns, jawlines, and fat deposition arrangement on the buttocks. It seems crisp and clear-cut: Primary are all those features that are directly involved in making babies. Secondary are all the other ways in which males and females—for various reasons—tend to differ.

The eighteenth-century Scottish surgeon John Hunter, who first coined the terms "primary" and "secondary" sex differences, did not have any qualms about the distinction either. But Charles Darwin, writing almost a century later, did. In *The Descent of Man, and Selection in Relation to Sex*, published in 1871, Darwin mused about the fact that when one tries to generalize across all animals, it becomes problematic to draw a clear line between primary and secondary: "[Secondary sexual characteristics] are not directly connected with the act of reproduction; for instance, in the male possessing certain organs of sense or locomotion, of which the female is quite destitute, or in having them more highly developed, in order that he may readily find or reach her; or again, in the male having special organs of prehension so as to hold her securely." So far no problems there. But then he went on to say that "[t]hese latter organs of infinitely diversified kinds graduate into, and in some cases can hardly be distinguished from, those which are commonly ranked as primary."

To understand Darwin's predicament, imagine the drumstick-like appendages on either side of the penis of the ladybird beetle (ladybug) *Cycloneda sanguinea* that it uses to tap the female during mating. Or the bright turquoise testicles of male l'Hoest monkeys. The ladybird penis and the mammal scrotum are supposedly primary sexual characteristics, but they have properties that seem unnecessary for transferring sperm. Unless we carry the distinction to its logical conclusion, Darwin wrote, and consider only the ovaries and testicles primary, "it is scarcely possible to decide . . . which ought to be called primary and which secondary."

In the end, Darwin avoided this gray area (probably much to the relief of his Victorian contemporaries; see Chapter 3) by staying well away from the genitals, stating that his book would chiefly be concerned with "sexual differences quite unconnected with the primary reproductive organs." He then duly proceeded to investigate the evolution of all kinds of body decoration, ornaments, and armature that male animals are adorned with but females aren't. Still, he did give us a way out of the difficulty of deciding between primary and secondary: by pointing out the distinction between evolution by natural selection and evolution by sexual selection.

Rhinoceros beetle horns, crustacean claws, deer antlers and prongs, stag beetle jaws, cricket and grasshopper song, bird plumage, and a whole range of other animal traits that distinctly differ between males and females are due to an evolutionary process that Darwin called sexual selection (or, in the title of his book, "selection in relation to sex"). In many ways, the discovery of this process was every bit as revolutionary as his discovery of evolution by natural selection, the focus of his more famous book, *On the Origin of Species.* We will return to Darwin and the theory of sexual selection in a later chapter, but for now let us focus on the fundamental difference between these two kinds of selection.

Evolution by natural selection needs four things to take place. First, there has to be variation between different individuals of the same species—say, in the numbers and sizes of fawn and maroon patches on a partridge's back. Second, this variation must be heritable; the offspring of a partridge with particularly large fawn patches on its back must also get relatively large fawn patches. Third, more offspring are produced than can survive. This is usually the case—a partridge lays up to twenty eggs; if all the chicks grew up, within a few decades we'd be knee-deep in partridges. The fact that partridges usually are much thinner on the ground means that most chicks do not survive into adulthood—they die of disease and are eaten by birds of prey. And the fourth condition is that death is not random—if the birds with more fawn on their backs are slightly less likely to be noticed by passing hawks in the dry grass in which they live, then the more fawn-colored partridges will have a slightly lower chance of dying than the more maroon ones. If all four of these conditions are fulfilled, then the stage is set for evolution by natural selection: a maroon species of partridge will, through natural selection by foraging birds of prey, and over many bird generations, evolve into a fawn species. It's a law of nature.

Sexual selection is different. Here, the great selector is not some extraneous entity like hungry birds, or parasites, or the weather. It is the other sex of the same species. If the environment, including partridge-eating hawks, did not favor one color over the other, the species would still evolve if partridge females preferred to mate with fawn-backed males rather than with maroon-backed ones (or even vice versa). Fawn-backed males would mate earlier or more often, with more different females and be able to father more chicks than maroon-backed ones, and sexual selection would be ongoing. Both sexual selection and natural selection boil down to the same thing—more of your genes in the next generation's gene pool—but they work by different means.

Having clearly demarcated sexual from natural selection, Darwin then returned to the problem of deciding which sexual characteristics are primary and which are secondary by making only one kind of selection responsible for each. *Primary* sexual organs, he said, are those that are maintained by *natural* selection. A male partridge needs organs to produce sperm and a thingy to squirt his sperm into the female. Similarly, females need the machinery to produce eggs and the plumbing to receive, store, and transport sperm. All these characteristics have been shaped by natural selection imposed by the bare necessities of life: individuals lacking any of these traits simply did not leave any offspring. But if partridge males have bright fawn backs, or red wattles under their eyes, or strange tufts of feathers on their necks, whereas females don't, then those *secondary* sexual characteristics are likely to have evolved through *sexual* selection, the result of the greater success that thus adorned ancestral males had over plain ones in the sexual strife over females.

Eminent biologist and philosopher Michael Ghiselin of the California Academy of Sciences in San Francisco has delved a bit deeper into the definitions of the terms we use when speaking of sexual characteristics—or sexual "characters," as biologists prefer to call them. I mentioned the human scrotum as well as the blue scrotum of other mammals as sexual characters. As Ghiselin has rightly pointed out, use of the term "character" is unforgivably sloppy. If having a scrotum is already a character, then having a *blue* scrotum cannot be a different character. Instead, Ghiselin thinks, it would be better to speak of "parts" on the one hand and their "attributes" or "properties" on the other. A scrotum is a part, but its color, whether naturel, bright blue, or bright pink as in the rhesus macaque, is an *attribute* of that part.

In this book, we will see that combining Ghiselin's reasoning with Darwin's is the best recipe for dealing with the confusing categories of primary and secondary sexual characters. Genitalia are primary sexual characters:

penises and vulvas and their multiform equivalents throughout the animal kingdom are "primary" because they are necessary "parts" that have evolved by natural selection. But most of their attributes—whether a penis is straight, coiled, two pronged, spined, double, or spatulate, for example—are the result of sexual selection and thus are secondary characters. In other words, most primary sexual characters are primarily secondary in character!

How to Be a Private Part

You probably realize by now that any distinction between primary and secondary sexual characteristics is a semantic morass. You will not encounter these terms anymore in this book. Instead, I will speak of genitals, genitalia, or genital organs. We may be jumping from the frying pan into the fire, though, because these terms still require definition. Fortunately Bill Eberhard, whom we already encountered in the Preliminaries, and who will grace these pages with recurrent appearances, has given us such a definition. Male genitalia, Eberhard said, are "all male structures that are inserted in the female or that hold her near her gonopore during sperm transfer." "Gonopore" is just a fancy word for vagina (which itself is a fancy word for a whole lot of other terms), and "near" is admittedly a little vague, but for the moment we have a good way to describe the territory of this book, as far as the male is concerned.

As for female genitalia, Eberhard stated: "I will consider as genitalia those parts of the female reproductive tract that make direct contact with male genitalia or male products (sperm, spermatophores) during or immediately after copulation." Again, there's some space for multiple interpretations there (what's "immediately after"?), but for our purpose Eberhard's definition of female genitalia will do fine. So, in a nutshell, in this book I will deal with the male machinery that transfers ejaculate to the female, and those female parts that receive and store it. (That also means I won't say very much about ovaries and testes, the organs that produce the sperm and eggs.)

It is important to realize that, thus defined, genitalia are organs that are present only in a limited set of animals, namely those that do internal fertilization. The myriad of waterborne creatures that simply shed their sperm and eggs into the waves do not have genitalia (and won't feature in this book). Again, we, from our human standpoint, are easily fooled into thinking that those "broadcast spawners" are the odd ones out. But in fact, it is we and all other landlubber animals that are really the weirdos here.

After all, animals evolved in the sea. For hundreds of millions of years, all the major evolutionary acts in the play of life had already been played out against a marine backdrop before, in the final act, a few twigs of that great evolutionary tree began spreading out on land: some plants, of course, fungi, some arthropods, snails, a couple of kinds of worm, vertebrates. The rest of life stayed safely in the briny womb of the sea. And how right they were to do so: marine organisms live in an environment that is extremely friendly to their sex cells. The saline solution chemically cushions their sperm and eggs; it is wet and has the same concentration of salts as do these cells themselves, so many marine animals can safely fertilize one another from a (great) distance by releasing into the currents their sperm, and often also their eggs, and trust that these will reach one another.

The situation faced by sex cells as they left the bodies of those first colonists on land, on the other hand, must have been like a Normandy beach on D-day. Spawning on land is out of the question: sperm and eggs will dry and shrivel and die in a matter of seconds. Even freshwater is deadly: unlike most cells, sperm cells cannot regulate the concentration of their salts, and as was first discovered by Dutch scientist Antoni van Leeuwenhoek in 1678, a sperm cell, when dropped in freshwater, will automatically imbibe so much water that it explodes in a matter of seconds. (This, incidentally, should lay to rest all those urban legends about women getting pregnant from previous guests' sperm clinging to the rims of hotel bathtubs.)

No wonder, then, that land and freshwater animals (and, admittedly, some marine animals—but for different reasons) have had to evolve ways to protect sperm during their trip from male body to egg. A fail-safe way is, of course, never to allow the egg to leave the body of the female, and to inject the sperm directly into the female body. And that is precisely what copulation and genitalia achieve.

Still, biodiversity being what it is, lots of animals that engage in what can only be called copulation choose not to use their penises to insert the sperm or their vaginas to receive it. Instead, they use parts of their body originally intended for a different purpose. Take rhodacarids, tiny soil-dwelling predatory mites. A male rhodacarid uses his jaws, not his penis, to transfer his sperm to his mate, which she may then absorb not with her vagina but through a pore on the base of one of her legs. For all intents and purposes, the male jaw and the female hip pore are their genitalia.

And mites are not alone in forgoing conventional sex organs in favor of a substitute. Before courting a female, a male spider fashions a special tiny sperm web, then "masturbates" into it and sucks up the sperm in his elaborate, fountain-pen-filler pedipalps—stubby arms on both sides of the head with hollow "boxing gloves" at the end. Then, pedipalps loaded, he wanders off in search of a female to woo and donate his sperm to. Although the sperm is produced by a pore in his abdomen, the business ends of his sex act (so, his genitals) are in his pedipalps. (Next time you watch a *Spider-Man* movie, imagine a more realistic substance shooting from those gloves.)

Using such replacement genitalia occurs quite a lot in spiders, mites, crustaceans, millipedes, dragonflies, and damselflies, and several other kinds of animals. Frankly, we don't really understand why or how this came about. Many of those still retain their original genitalia but have stopped using them as such, promoting other body parts to that position. And to end this chapter with a bang, I will give you two stories of animals that have taken substitute genitalia to new heights: cephalopods—the group of mollusks that includes octopi, squid, and cuttlefish—and velvet worms.

Story 1: Calamari Coition

In June 2012, one of those strange-but-true news items rippled across the world to be grossed out at briefly and then forgotten: "Woman, 63, Becomes PREGNANT in the Mouth with Baby Squid After Eating Calamari" was one of many similar headlines in the newspapers. What had happened?

Immediately after eating a mouthful of parboiled squid at a South Korean seafood restaurant, a customer had been rushed to the doctor with "severe pain in her oral cavity." There, twelve squid sperm packages, or spermatophores, were found embedded in her tongue, inner cheek, and gums. Apparently, as a posthumous attempt at cephalopod-human hybridization, the (male) squid that she had eaten had ejaculated a bunch of his spermatophores into her.

The medical experts studying this case (and newspaper readers across the world) were stunned, but to somebody familiar with squid reproduction, it came as less of a surprise. José Eduardo Marian of the University of São Paolo in Brazil is one such expert, and as he writes in a 2012 paper in the journal *Zoomorphology*, there are in fact at least sixteen similar cases in the medical literature, most from Korea and Japan, where eating raw or blanched squid is common. To Marian, that a dead male squid can still ejaculate and have his sperm packages lodge themselves into a person's mouth is not unexpected, given that these spermatophores, as he writes, "function autonomously and extra-corporeally."

Squid spermatophores are intricately constructed flask-like things of less than a millimeter (0.04 inch) ranging up to tens of centimeters (about 10 inches) long, the latter being found only in giant squid. Composed of several layers of membranes (some of which are tightly coiled and under tension), a bag of sperm, sticky material, and, in certain species, abrasive spikes, they are nothing less than spring-loaded sperm grenades. All the male does is ejaculate them and then deposit them in or on the body of the female. Once there, either triggered by the sea water or by the friction of leaving the male's penis, the spermatophores self-ejaculate—yes, the ejaculate can ejaculate! The outer membrane bursts, the spring unleashes itself, the abrasive spikes cut a hole, and the sticky cement helps to secure the payload to the female's skin. Or to a diner's oral epithelium, for that matter. In fact, as Marian points out, much of what is known of the way squid spermatophores function is thanks to the steady supply of samples embedded in human tissue collected by Japanese and Korean emergency rooms.

The point is: the female squid does not have a vagina as such. Depending on the species, the male may deposit his spermatophores around her mouth, on her back, her arms, or inside her mantle—the rubbery mitten-shaped covering of a squid or cuttlefish that we humans cut up in rings, deep-fry, and eat as calamari. From there, when the eggs are laid, the sperm find their way to them on their own, although the females of some species store sperm in special pockets around the places where the spermatophores are normally stuck.

And not only do squid females lack a vagina, squid males lack a penis. Or

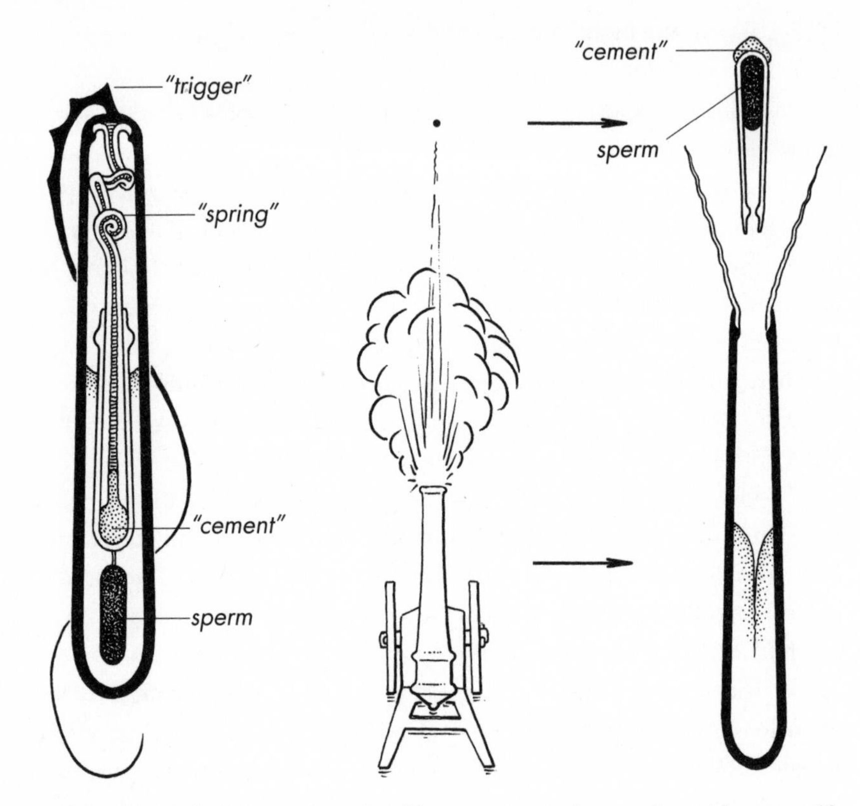

A spring-loaded sperm grenade. The sperm package of squid can self-detonate, which causes the sperm payload to be propelled and stuck to the body of the female.

rather, they do have a penis, but they don't use it as one. Let me elaborate. Most squid eaten in restaurants around the world belong to species in which the males have only a very short penis, too short to reach out of the mantle opening underneath the squid's "neck." So how do they manage to place spermatophores so precisely on, for example, the rim around a female's mouth? For this, they normally use one of their eight arms, one called the hectocotylus, which is specially adapted for handling spermatophores. (It often has a special groove and folds, and lacks suckers along part of it.)

During mating, when male and female squid are locked head-to-head, arms grappling in bliss, the male reaches with his hectocotylus inside his own mantle and produces the spermatophores, which his penis has just released, and places these in or on the female. So, in fact, and according to Eberhard's definition, the hectocotylus—not the penis—is the actual genital organ, since it is what delivers the sperm to the female.

In one special group of cephalopods called argonauts, the hectocotylus even takes on a life of its own. Argonauts, or paper nautiluses, are mysterious octopi that live virtually unstudied lives in the open ocean. From what little we do know about them, it is clear that the seven species of argonauts are in many ways among the strangest of cephalopods. To begin with, the females—translucent, purplish, and bluish spotted—are octopus-sized, but the males are tiny: usually just 1 or 2 centimeters (0.4–0.8 inch) long. Also, the females have two webbed arms that are so large that the great Linnaeus, who named the first argonaut species, *Argonauta argo,* thought they held them up above the water surface as sails (hence the name, after the mythical ship *Argo* and its sailors). Not so: the female uses these arms to fashion a paper-thin shell that is the spitting image of the kind of shells that were once produced by the extinct ammonites. And herein lies yet another argonaut oddity: in contrast to all other mollusks, the argonaut shell is not attached to the body. Instead, the female lays her eggs in it, keeps it buoyant with bubbles of air, and guards it until her babies have matured.

But before eggs are laid, copulation needs to take place. And argonauts' copulatory habits are unorthodox even for a cephalopod. A male argonaut has a very large hectocotylus, to which it attaches his spermatophore. That, as we have seen, is nothing out of the ordinary in the cephalopod world. But the male argonaut mates only once in his entire life. He has no choice, because during mating he detaches his entire hectocotylus, which then autonomically wiggles its way into the female's mantle cavity and stays there until she lays her eggs. Sometimes, a female hosts several of such live hectocotyli, from multiple sexual encounters. In fact, the argonaut hectocotylus is so strange looking, with a head and a tail, that the French zoologist Georges Cuvier in 1829 mistook it for a parasitic worm and gave it the scientific name *Hectocotyle octopodis.* Although later scientists discovered its true nature, the name stuck, and hectocotylus is now the name used for the male sexual arm—in effect, his genitalia—in all cephalopods. And it is likely that that customer at the South Korean seafood restaurant had already swallowed her squid's hectocotylus before her mouth was impregnated by his spermatophores.

Story 2: Blue Velvet

Back in 1883, one hundred pounds sterling was a lot of money, especially when spent on a grayish worm-like creature living out in the South African Cape. And yet that is what the British Royal Society paid zoologist Mr. Adam Sedgwick to seek out the animal, then known as Peripatus, and bring it back to England for study of its reproduction. They had good reason to want to study these creatures. Since their discovery in the 1820s, the mysterious Onychophora, or velvet worms, as we now call them, had fired the imagination of zoologists. With their thin, soft, knobbly skin of chitin and their system of tubes to transport oxygen through the body, they resembled insects and their relatives, but their numerous stubby legs and the rings that divide their 2-to-10-centimeter-long (1-to-4-inch-long) bodies made them look more like annelid worms. Even more fascinating was the fact that the females give birth to live young, which they gestate for more than a year in something that suspiciously resembled the uterus of mammals, placenta and all.

We now know much more about onychophorans. We know that they are a separate phylum of animals, equal in rank to the arthropods, and indeed closely related to them (but not to the annelids). We know that there are hundreds of species living in damp places all over the Tropics and the Southern Hemisphere, where they (sometimes in family groups) hunt termites and other prey, which they catch with a sticky substance squirted from nozzles under their mouths. We know that they have a surprising variety of ways of reproducing—some indeed have extended pregnancies after which they give birth to live young, but others lay eggs. And despite their rarity and elusiveness, they have become somewhat of an icon, due to a nonscientific characteristic known as "cuteness." A commenter to one of the many velvet worm videos on YouTube likened them to "centipedes in romper suits." Quite.

But adorability aside, they figure in this chapter for the ways in which males transfer sperm to females. As noted by Adam Sedgwick, who brought three hundred live onychophorans from the Cape to Cambridge and studied them there, "the males deposit spermatofors quite casually all over the body of the female." He added: "How the spermatozoa pass up the uterus and oviducts, which are always full of embryos . . . I do not know."

In the nearly 130 years that have passed, we have learned a great deal more about onychophoran sex. And it is weirder than Sedgwick could have dreamed.

Although some velvet worms have orthodox vaginal sex, males in the family Peripatopsidae, to which Sedgwick's species belonged, produce spermatophores that they, not unlike squid, affix to the female's skin. Sedgwick was rightly puzzled by how the sperm then reach the eggs, given that they would first need to find their way to the vagina and then struggle through a uterus cluttered with developing embryos. The fact is that they don't. Instead, the female helps them create a rather drastic shortcut. At the site where the sperm package sticks to her skin, she releases enzymes that dissolve her own skin as well as the spermatophore envelope, which allows the lump of sperm to sink into her body. The sperm cells then begin swimming through the female's blood. They reach her sperm storage organs either via helpful funnel-like structures or by forcing themselves straight through the wall of these organs.

But it gets stranger still. In the 1980s, velvet worm specialist Noel Tait of Macquarie University in Sydney discovered a whole menagerie of unknown species, all belonging to entirely new genera, in the forests of New South Wales and Queensland, of which the males all carried curious "head structures" between their antennae. These ranged from flower-like arrangements of bumps to single or multiple erectable teeth and spikes. Only in 1994 did it become clear what the males use these for. On April 27 of that year, Tait found a mating pair of the species *Florelliceps stutchburyae* in leaf litter in the Nightcap Range, a mountainous area some 130 kilometers south of Brisbane. As Tait described it in the *Journal of Zoology*, "The male's head was firmly attached to the genital region between the last pair of [legs] of the female. The two individuals moved about in a coordinated fashion with the female leading." When he took a closer look at the couple under the microscope, he saw what was happening: the male had a spermatophore stuck on the prong on its head, which he was delivering into the female's vagina. The female, meanwhile, had her hind claws locked tightly into the skin around his head structure, as if to keep him in position.

For these male velvet worms, it seems, the crowns on their heads are their genitalia. These are the tools they use to move sperm into the female's vagina. A further twist came in 2006, when Tait and his collaborators

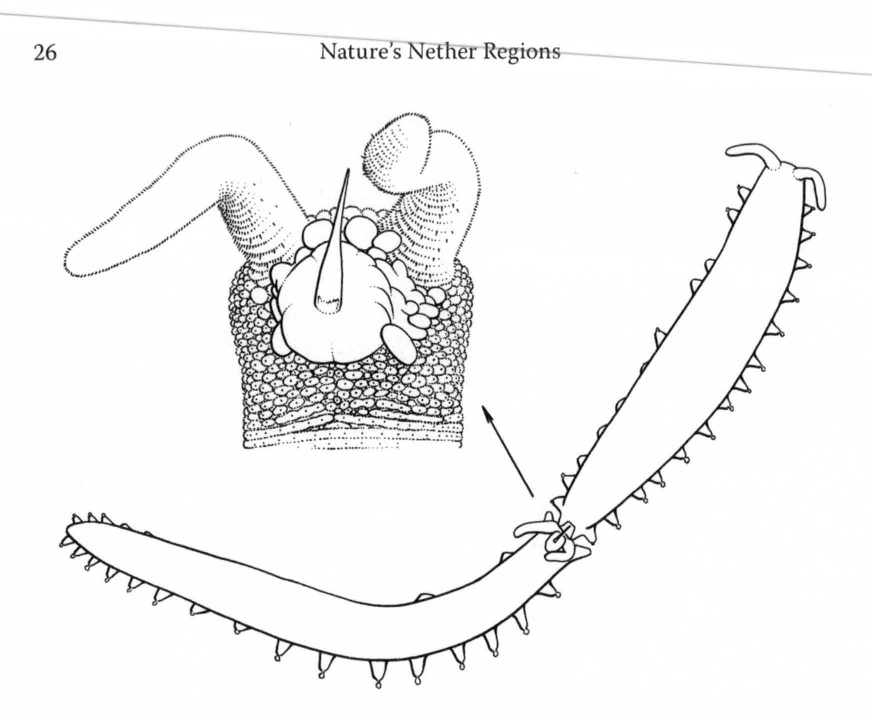

Genitalia on your head. The male Australian velvet worm *Florelliceps stutchburyae* mates by placing sperm on a crown-like adornment of his head and then pushing his head into the female's vagina.

announced that they had found that females of several head-structure species still possess working machinery for transporting sperm from their skin to their eggs via the blood. So Tait now thinks that perhaps the head-first mating behavior is applied only on virgin females, which still have a vacant uterus, not clogged by developing embryos, whereas mature females, with developing embryos in place, mate with the old through-the-skin method. Whether the males also use their head structure to fix spermatophores on the female's outside remains to be discovered. Either way, it is clear that velvet worms give a whole new meaning to the term "genitalia."

At the end of this first chapter, we are left with the sobering thought that in sex, all our certainties should be reconsidered. Even if the immediate reason for a male and a female to have sex may be attraction, hormones, and the innate urge to procreate, the ultimate cause is an unlikely sequence of evolutionary events involving genetic variation to evade parasites, peacekeeping among warring organelles, and the hostile environment out of the sea. The tools they use to accomplish all this, the organs subsumed under the term "genitalia," even if we adhere strictly to Eberhard's definitions, can be anything from a penis to a zombie tentacle and from a vulva to an upper lip.

Yet it is precisely those genitalia that evolution seems to have singled out to display its greatest virtuosity. Next we will see that taxonomists, not evolutionary biologists, were the ones with front-row seats to this evolution tour de force.

Chapter 2

Darwin's Peep Show

I have now traversed the dusty paths crisscrossing the Jardin des Plantes and the buildings of the Muséum National d'Histoire Naturelle. I have leafed through *Les Statues du Jardin des Plantes* in the museum bookshop and scanned the numbered, colored dots on the plans at the entrance of the famous Parisian park. Arrows point to the grandiose statues of the great French naturalists Lamarck and Buffon, and the book provides directions to at least twenty-five other busts in the gardens. But the one I am after, of entomologist René Jeannel, remains elusive.

For the third time, I cross rue Buffon and wander among the haphazardly placed nineteenth- and early twentieth-century buildings, in various states of disrepair, that form the small village of zoological and geological laboratories of the museum. The security guard, in his cabin at the main gate, who has already been ogling me suspiciously, says he has no recollection of a statue anywhere nearby. But then a second security guard who ambles in overhears my question and relieves my two-hour-long foray with the words: "Ah, a statue? Yes, there is a statue at the end of this cul-de-sac, in the corner, behind that blue truck!"

Indeed there is. The truck obscures from view an ill-tended corner of the grounds of the entomology department, where building rubble, a garden shed, and parked cars vie for space. Among them, barely rising above weeds and saplings, and with several wooden pallets propped up against it, stands the bust of a solemn mustachioed gentleman, a gift from the Romanian government on the passing of Dr. René Jeannel (1879–1965), "for his magnificent oeuvre in entomology, biospeleology, zoogeography, and evolution."

Judging from photographs, Jeannel in real life looked more kindly and well coiffed than the wild-eyed, ominous effigy that the Romanian artist, probably steeped in socialist realism and portraying revolutionary heroes, had produced. Notwithstanding representational inaccuracies, the bust and its subject deserve more exposure than they get in their forgotten corner of the Paris museum grounds. For Jeannel, who was indeed a great entomologist, was also one of the founding fathers of the study of genitalia.

The son of a medical officer in the French army, Jeannel was destined to follow in his father's footsteps. But his medical career had already been nipped in the bud when, during his studies in Toulouse, at the foot of the Pyrenees, Jeannel took up caving and began exploring the many limestone caves in the south of France. Initially as a hobby but soon more and more seriously, he studied the bizarre life-forms adapted to the dark, damp, cool, and nutrient-poor conditions deep inside those ancient caves. During those excursions, Jeannel discovered two unknown cave beetles in the Grotte d'Oxibar, which an entomological authority published for him and named *Bathyscia Jeanneli* and *Aphaenops Jeanneli*. By then, any residual medical aspirations were doomed, much to his father's regret: René was going to be a full-time naturalist.

In 1905 Jeannel teamed up with another young spelunker, the Romanian biologist Émile Racovitza, and together they embarked on a lifelong collaboration in biospeleology: the study of the evolution of "troglobites," those bizarre organisms that have adapted to the demands of cave life and can no longer survive aboveground. The two must have almost grown into troglodytes themselves, spending more time in those underground echoing vaults than in the human world, for in the first seventeen years of their collaboration alone they jointly explored a staggering fourteen hundred caves in southern Europe and North Africa. They published descriptions of the caves and the animals living there, and in passing discovered some of the finest examples of Paleolithic cave art. Later, when Racovitza was invited by the government of his home country to found a biospeleological institute in Cluj, Jeannel joined him there as deputy director (hence the Romanian statue) until he landed a job at the Paris museum in 1927, accepting a chair in entomology and eventually becoming the museum's director.

Despite these managerial posts, Jeannel was primarily a hardworking taxonomist who would closet himself into his office for days on end whenever he was working on a problem. By 1911, which saw the appearance of his 641-page, 657-illustration doctoral thesis on a band of cave beetles called Leptodirini, he had already published more than thirty papers. And during the rest of his working life he added another five hundred scientific publications to this total, together amounting to more than twenty thousand published pages, mostly on cave insects and all illustrated by his own deft hand.

The cave beetles that Jeannel studied are very old: derived from an aboveground ancestor that must have lived in the Mediterranean tens of millions of years ago, they colonized the many isolated caverns as these were gouged out by groundwater and subterranean rivers, and have had ample time to adapt to the unique cave environments. Pale, with long, spindly legs and ditto antennae that make up for the loss of eyes and wings, as well as bloated abdomens that house extended guts for extracting all the nutrients they can from what little food enters the caves, they all pretty much look alike. And yet, as Jeannel and his colleagues discovered, there are thousands of different species, many of which occur only in a single cave system, evolved in isolation from relatives living in the cave system next door. With their external appearance so constrained by the requirements of their environment, the fact that these are all different species is not nearly as apparent in their outward form as it is in their internal organs—in their genitalia, to be precise.

Like many beetles, a leptodirin cave beetle has a penis with, on either side, a whip-like "paramere" with a couple of bristles at the end. The penis itself is hollow and contains a soft, crumpled-up sac that, during mating, is blown up and extrudes via two flaps at the bottom. When fully inflated, the internal sac turns out not to be as soft as it appeared at first: it is studded with rows of tough teeth, larger spines, and sometimes a few extremely long and sharp spikes. Jeannel did not (as we will do in Chapter 7) stop to wonder about the function of this entire rather vicious-looking contraption. Instead, for each species of leptodirin, he meticulously described the shapes of the parameres and the arrangements of the bristles at their top, the shape and curvature of the penis, and especially the exact way the internal sac was adorned with rows of variously sized and shaped spines. In so doing, he revealed that hundreds of variations on this theme exist, corresponding to just as many different species. Not only that, he also discovered that the way the male genitalia are constructed holds the key to species' classification, and he was the first to group species into families based on how their penis looked.

The beetle penis became Jeannel's bread and butter. Aware of its potential, he invariably assigned projects to his students that involved dissecting, describing, and categorizing the penises of beetles and other insects, and was puzzled when, in one case, this brought a female student to tears. She told him that he could not possibly expect such a thing of a lady. In what must have been an effort of empathy, Jeannel defused the situation by pointing out that he was not actually asking her to study penises; instead he preferred to use the word "aedeagus" (from the Greek *ta aidoia,* "the genitals"), since "penis," "phallus," and "prepuce" are terms usually reserved for vertebrates like ourselves.

A Plethora of Parts

In the early twentieth century, recognizing and identifying animal species by their aedeagus was still something of a novelty among taxonomists, and Jeannel was certainly a trendsetter in his time. Where one of Jeannel's predecessors, the vertebrate paleontologist Cuvier, had as his adage "Show me your teeth and I will tell you who you are," Jeannel's might well have been "Show me your willy. . . ." Toward the end of his career, in 1955, he devoted his entire 155-page memoir *L'Édéage* (the Frenchified version of the word) to the insights the aedeagus had given him. Today, however, biologists routinely use male and female genitalia as a quick and easy way of distinguishing species that often are very similar otherwise. Or ones that are so variable in color or size that the shape of the reproductive organs is the only reliable indicator of a species.

Bumblebees are a good example. Watching the large, furry insects in their jolly colorful outfits heave themselves from flower to flower, you might think that, insect field guide in hand, one could easily determine the name of a species by a quick scan of its conspicuous black, yellow, and red pattern. Unfortunately, the reality is more frustrating. In Kashmir, for example, some thirty species of the genus *Bombus* exist, each of which seems to own roughly the same wardrobe of woolly black with variable yellow, orange, and red cross-band patterns. As a result, the color of a cashmere bumblebee says near to nothing about its identity. Experts who are in the know can discern the subtle species-specific differences in the shape of the jaws and the antennae, and in the pattern of pits and wrinkles on the surface of the head. Still, the only fail-safe way of identifying many of these species is to get your hands on a drone, extract his penis, and compare its shape with pictures of the characteristically shaped penises of all thirty species. And it's the same with the twenty-four species that occur in Britain, or the thirty-five or so North American bumblebees.

I hope that with these entomological examples, I have not given you the impression that this genital pinnacle of biodiversity is something specific to tiny creepy-crawlies. In fact, the pattern is pervasive throughout the whole animal kingdom, right up to and including mammals. Take, for example, the eighteen species of elephant shrew. These insect-eating animals, which live in all kinds of habitats across the African continent, range from mouse-sized to opossum-sized. Apart from their somewhat long legs, their extended noses, and their sometimes colorful or checkered fur, elephant shrews resemble true shrews, with which they were classified in the past. DNA studies, however, have revealed that elephant shrews are actually members of the Afrotheria, an ancient group of predominantly African mammals that also includes aardvarks, elephants, manatees, tenrecs, and hyraxes, among others.

In addition to the funny trunk-like nozzle that their noses have evolved into, elephant shrews have peculiar genitalia. Females lack a vagina as such (instead, the womb opens directly to the outside world) and are, besides primates and bats, the only mammals that menstruate. And the males have their testicles as well as most of their penis hidden inside their bellies. The elephant shrew penis is unusual in other ways, too. It is extremely long—about half the animal's body length—running along the inner wall of the belly all the way from his hindquarters until near the breastbone. There it makes a sharp U-turn and emerges from the belly pointing downward and backward—although in erection, muscles in the skin direct it forward again, so that the shaft of the penis becomes, when seen from the side, Z-shaped.

If we were to follow an elephant shrew's penis from its base to where it emerges from his furry belly, we would see that all elephant shrews follow the same basic design. But it is when we reach the tip of the penis that all hell breaks loose. Peter Woodall, whose in-depth study of the elephant

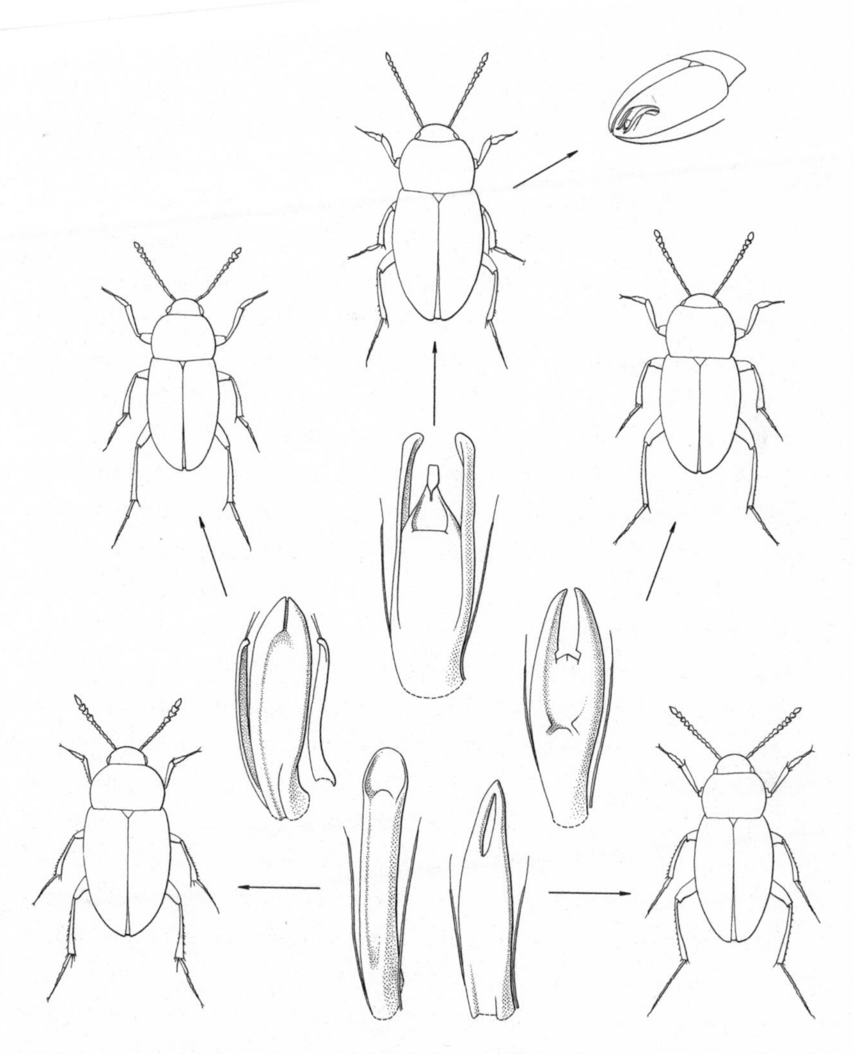

Show me your willy. Animal species that are near identical externally often have very different genitalia. In this picture, five—very similar-looking—*Catops* beetle species are shown, together with their respective—very different—penises. In life, the penises are hidden in their abdomens and not visible on the outside.

....

shrew's penis tip (don't laugh) appeared in the November 1995 issue of the *Journal of Zoology,* explains that the golden-rumped elephant shrew, from Kenya, has a row of spines on the tip, which ends spoon-shaped; the Southeast African four-toed elephant shrew has a sharp tip with two sideways-pointing "ears," like a medieval ranseur; and the South African short-eared elephant shrew has a collar near the end and a puffed-up tip; whereas all species that belong to the genus *Elephantulus* have a nub at the end of the penis tip that can be heart-, dish-, boomerang-, or flower-shaped, depending on the species.

I could go on like this, but I won't—at least not yet. You will have to take my word for it that we find this pattern in almost any kind of animal we turn to—to such an extent that it is almost a law of nature: of all the organs that an animal is provided with, the greatest differences between species are not in their brains or beaks, or in their kidneys or guts, but in their genitals. This applies to cave beetles, bumblebees, and elephant shrews, as well as to velvet worms, land slugs, water and rove beetles, small ermine moths, daddy longlegs spiders, banana and hover flies, egg parasitoid wasps, aquatic annelid worms, hoofed mammals, sharks and rays, primates, guppy fish, damselflies, land planarians, nematode worms, trombidiform mites, and harvestmen. To name but a few.

Taxonomists never stop feeling privileged for being the ones to explore and discover firsthand this amazing diversity of form and shape. To them, dissecting out the genitalia of previously unstudied specimens is like unwrapping Christmas presents; never knowing what unexpected shapes will come to light, their concentrated, painstaking anatomical efforts behind the microscope are punctuated by muffled cries of delight. When I was still an impressionable grammar school boy and began frequenting the local meetings of the Netherlands Entomological Society (held in that deer antler lecture theater described in the Preliminaries), the first seminar I attended featured a famous Russian moth specialist. Frail and aging yet with such a youthful enthusiasm he could barely contain his glee, he told his audience how he had just finished studying a box of tiny pinned specimens collected in a faraway, previously unvisited corner of the Asian continent, home to lots of species new to science, and had been amazed by the genitals he discovered. Flashing some slides of shapely slivers of chitin covered in hairs and spines, he exclaimed in exasperation: "This. Is. Totally. Unheard of. And unseen!"

To fully reveal the significance of this hidden beauty of sexual shapes, their sensuous implications, and the sheer endlessness of the terra incognita of forms requires not just science, but also art. Colombian-born artist

Maria Fernanda Cardoso managed to pull this off with her project *The Museum of Copulatory Organs.* Throughout her career, Cardoso has used living and dead animal and plant parts to create mesmerizing sculptures from bones, dried reptiles, and butterfly wings. At the fifty-second Venice Biennale she represented Colombia with a large piece made out of dried starfish. And in the 1990s she gained international fame with her installation *Cardoso Flea Circus,* an artistic interpretation of the nineteenth-century circus sideshow performance.

While researching for the flea circus, she came across a startling reference to the complexity and size of their genitalia. When the flea circus project was over, she decided to find out "if fleas were the only well-endowed arthropods or whether there were more." Now based in Sydney, Australia, Cardoso spent a summer in the library of the Australian Museum, soon hitting upon Bill Eberhard's book *Sexual Selection and Animal Genitalia* and realizing how much potential there was for this project. Collaborating with entomologists, 3-D modelers, and electron microscopists, she has since produced series of glass renderings of pseudoscorpion spermatophores, damselfly penises in bronze, and large-scale blowups of snail genitals.

Most evocative are her representations of the penises of nine species of harvestmen from Tasmania. Starting from electron micrographs, Cardoso first had these fantastic shapes transformed into three-dimensional computer models and then printed in white plastic with a 3-D printer. Finally, she placed them under tight-fitting bell jars. As Elizabeth Ann Macgregor, director of the Museum of Contemporary Art Australia, said in a TV documentary on Cardoso's work: "You see these shapes, and of course you immediately think of a penis, because she's put them under these glass structures, which could be condoms. So immediately the references start. And then you begin to look closer. They're extraordinarily beautiful—they become like flowers."

Hand in Glove

Of course, the pressing question that this genital burlesque impels is, Why? Why would different species have completely different shapes for organs that perform the same simple function in all of them? Different species eat different food, so having different sets of teeth, as Cuvier emphasized, makes sense. But different penises and vaginas? For a long time, biologists smugly leaned back in their chairs, claiming they knew *exactly* what all this genital diversity meant. Even before Jeannel, deep in the nineteenth century, the naturalist Philip Henry Gosse already wrote about this: "If I see a number of keys, of very minute and elaborate workmanship, all different, I cannot doubt that every one is intended to fit some special lock."

The idea is that each species is provided with its own unique penis and vagina combination to assure that only members of the same species can copulate successfully. At the same time, crossbreeding with other species is barred (either because there is no fit or because it "doesn't feel right")—hence the aptly named "lock-and-key" hypothesis. Intuitive and aesthetically pleasing, the idea makes perfect sense. To begin with, hybridization between different species, when it happens, often leads to offspring that are sterile (think of mules), sickly, or poorly adapted, so it stands to reason that evolution would make sure that such poor bastards were prevented from being born. Furthermore, wasting time and, worse even, sperm and eggs on mates belonging to the wrong species could be prevented by a system of keys that don't fit into other species' locks. And finally, nature is full of situations in which there is a real risk of such sexual mistakes; closely related species often share the same habitat.

A little foray into necrophagy illustrates this last point. If, like ecologist Petr Kočárek did, you were to take a close look (hold your nose!) at the beetles living on carrion in a Czech forest, you'd find that closely related cadaver-feeding species are rubbing shoulders with one another. Kočárek found up to eleven species of *Catops* beetles living together on the same piece of rotting meat, up to nine species of the rove beetle *Atheta,* and five different *Nicrophorus* burying beetles. All these beetles are, at least to the human eye, very similar. Those *Catops* species, for example, are all about 3 millimeters (0.1 inch) long, gray, and oval, with pretty much identical legs, antennae, and body proportions. But each species has its own unique penis shape: *Catops tristis:* lance-shaped. *C. nigrita:* three-pronged. *C. morio:* gouge-like. *C. westi:* forked. *C. chrysomeloides:* spoon-shaped. It is not hard to imagine why a check for mutual fit of lock and key might be an efficient safeguard against accidentally having it off with the wrong species.

And as further support for the idea, biologists discovered that male and female genitalia of the same species indeed do mesh perfectly. In the May 27, 1967, issue of *Nature,* primatologist Jack Fooden of the Field Museum of Natural History in Chicago provided such an example for two monkeys from southern China: the rhesus macaque (*Macaca mulatta*) and the stump-tailed macaque (*Macaca arctoides*). The rhesus macaque has a human-like vagina and a blunt, helmet-shaped glans penis. The stump-tailed macaque, on the other hand, has its glans extended into a long, flat, lance-shaped structure of up to 7 centimeters (3 inches) long, supported by a bone inside. At the same time, the vagina of the female of the same species is almost completely obstructed by a thick lump of tissue hanging from the roof. As Fooden writes, the slender penis "is ideally formed for reaching and entering the vaginal opening of the female by passing through the narrow slit. . . . [C]onversely, it appears unlikely that this passage would transmit the short blunt glans penis of *Macaca mulatta.*" In other words, it would be like the proverbially doomed square peg in a round hole.

With so much going for it, it's little wonder that the lock-and-key hypothesis survived for more than a century without anybody seriously questioning it. Another reason may be that, until a few decades ago, the biologists most interested in genitalia were taxonomists, quietly organizing and dissecting, drawing and describing their specimens, and not particularly keen on rigorous tests and experiments. Yet dark clouds did appear on the horizon now and then. Already in the 1920s biologists working on bumblebees (then still called by the slightly more endearing "humblebees") realized that the multiformity of male "keys" (already showcased above) was answered with just a single "lock" in the females. Despite attempts to find differences, bumblebee researchers had to conclude that the vaginas of these winged balls of fluff all have the same internal shape. More worryingly still, with alarming regularity reports appeared of males of one species of insect found with their keys firmly and cheerfully inserted in other species' female locks. A butterfly collector in Rotterdam once showed me his drawer full of what he called "flagrante delicto cases": pinned female butterflies, the male of another species still hanging by its genitals from hers.

But it was not until the 1980s, when evolutionary biologists began to pay more than cursory attention to genitalia, that the theory found itself in stormy weather, and it has since then been slowly sinking to the bottom of the scientific sea. So how did such a simple and attractive idea that was quite popular for more than a hundred years all of a sudden fall out of favor? First of all, when you think about it (as some clever people started doing), the theory does not make as much sense as it seems to at first sight: if the prevention of miscegenation had indeed been the evolutionary drive, then wouldn't placing the crucial barriers at the *very end* of the sequence of preliminaries be a bit inefficient?

There was more. I already mentioned that, very early on, researchers found that the differently shaped penises in bumblebee males seem not to be matched by differences in the female vaginas; hence, the whole lock-and-key idea fails in these animals. In fact, with few exceptions, genital differences between species sit in the male parts, and much less in the female ones.

In a brief aside, I should stress that the apparent absence of female "locks" may be an illusion caused by an issue that will be a recurrent theme in these pages: the worrying ignorance of the ins and outs of female genitalia compared with the well-studied male ones. Zoology has not been spared the same male chauvinism that has made the study of the human male sexual apparatus deemed more acceptable and relevant than parallel studies in females. There is also a practical bias: male genitalia are often sturdy, sticking out, and anatomists find them much easier to access than the soft, folded, invaginated ones of females, especially when we are talking about dried and pinned millimeter-sized insects. Both the sociology of science and the realities of preservation and dissection, therefore, seem to have conspired in our ignorance of female genitalia. So in many animal groups where the female genitalia are said to be "all the same," this should be interpreted as "nobody has really bothered to look." We will come back to this at length, also in reference to human genitals.

With this caveat in mind, even in types of animals where scientists *have* taken the trouble to dissect and measure the female genitalia with as much abandon as the male ones, the general trend points to genital extravagance more in males than in females. At the same time, reports keep popping up of females unable to keep their fertilization apparatus locked to "wrong" males.

One particularly telling example of this last problem comes from

Ciulfina praying mantids. These small (as mantises go), handsome insects live in forests in Australia's northeast and are genitally quite special in that some species have male genitals that are the mirror image of those of other species. You'll have to imagine that the penis of a praying mantis is always a very asymmetric affair. Seen from the rear, it is a twisted arrangement of plates, spokes, and prongs. And as with all asymmetric forms, its mirror image cannot be superimposed on the original. This is called chirality, after the Greek *cheir,* meaning "hand": our hands are perfect examples of chiral forms, with your left hand identical and yet completely different in form from your right. (We will come back to chiral genitalia in Chapter 8.) So some *Ciulfina* species have, let's say, "right-handed" male genitals whereas others have "left-handed" ones. However, says Greg Holwell of the University of Auckland in New Zealand, who studies these critters, the female genitalia are all symmetric. The result is, as Holwell found out when he mated females of left-handed species with right-handed males and vice versa, that females' locks can be opened by their own males' set of keys just as easily as by a set of mirror-image keys. Try that with a real lock, and you'll understand that this is not something that jibes with a lock-and-key idea.

Still, this argumentation—the absence of female locks—could be seen as only circumstantial evidence against the lock-and-key hypothesis. After all, it is possible that genitalia do not work like a traditional lock and key but rather like one of those modern electronic key cards: no mechanical mesh but an imperceptible exchange of sensory signals, something that need not be visible in the female genitals' shapes.

No, the more damning verdict against the genital locksmith came in the third chapter of Bill Eberhard's *Sexual Selection and Animal Genitalia.* Rather than looking for proof (or the lack thereof) of a tight fit—either mechanical or sensory—between male and female genitals, his approach was, Let's look for situations where we would *not* expect the lock-and-key system to evolve. What do we see there? Such situations would be species that simply never run the risk of encountering an amorous member of a closely related species, and therefore would have no use for any specialized locks or keys. The leptodirin cave beetles that Jeannel studied, for example. Each cave has its own species of beetle that never ventures outside of it, and hence never, ever meets any members of the related species in the neighboring cave. Or think of island archipelagos, where each island often has its own endemic species that lives only there and nowhere else, to be replaced by a related, but different, species on the next island. And then there are parasites, which live and mate only on or in a particular host; to them, their host is their island, and they never meet the closely related brethren that are specialized on other hosts.

Eberhard trawled the scientific literature and came up with dozens of such cases: the *Oryzomys* mice of the Galápagos Islands, the checkerspot *Atlantea* butterflies of the West Indies, *Meropathus* water beetles of the sub-Atlantic islands, and also the lice that live in pocket gophers' fur, the feather lice of crows, the pinworms in the intestines of humans and other primates . . . all groups of multiple species each inhabiting its own "island" and never meeting another species of the same group. And yet all showed genital differences between species just as great as between species that *do* actually run a risk of mating with the wrong species. In the light of the lock-and-key hypothesis, this does not make sense. It would be as silly as the castaway sailor on an uninhabited island who locks and bolts his hut whenever he goes coconut collecting.

Since Eberhard's book, such evidence against the lock-and-key hypothesis has been accumulating. By now, it seems an open-and-shut case. Clearly, as we will see in the next chapter, we must look for other explanations for the genital extravagance that appears to be commonplace in nature. Still, I have to leave you with a paradoxical truth: the demise of the lock-and-key theory does not mean that there are no animals that are barred from mating with related species because of nonmatching genitals. We already hinted at this with the rhesus and stump-tailed macaques, but there are more.

A good example of a true lock-and-key situation was published in 2012 by evolutionary biologists Yoshitaka Kamimura and Hiroyuki Mitsumoto of Keio University in Japan. They studied *Drosophila santomea* and *D. yakuba,* two species of banana fly from the small African volcanic island of São Tomé. Males of the species *D. yakuba* carry two sharp spines on the base of the penis, which during copulation fit exactly in two reinforced pockets on the female's vagina. The other species, *D. santomea,* lacks both these spines, as well as the pockets. As a result, a mating between a male *D. yakuba* and a female *D. santomea* is no fun for either party: the female, lacking any protective armor on her vagina, is wounded by the spines on the male's penis. He, conversely, is not happy either, because the spine/pocket mismatch causes his penis to be misaligned, with the result that, when he ejaculates, his sperm does not enter the female's vagina, but instead is spilled on her behind. The physics at that spatial scale being what it is, the drop of semen dries instantly, gluing the pair (whose enthusiasm for each other was already rapidly dropping) firmly to each other. This then leads to a half-hour struggle during which male and female kick each other vigorously until they finally break loose.

Other such cases of lock-and-key mismatch in the genitals do exist. Still, these are not to be taken as evidence for the lock-and-key hypothesis. The point is, in none of these cases is it likely that avoiding crossbreeding was the *cause* for lock-and-key-like pairs of genitals to evolve. More likely, the inability to mate with other species was simply an unfortunate side effect of genital shape differences evolved for very different reasons. And those reasons are—as we shall see in the next chapter—much more titillating than locks and keys.

Chapter 3

An Internal Courtship Device

In a small room in the attic of my home I have created my own explorer's den. Nestled among the roof beams is a jumble of wooden and rattan furniture, a shelf with shells, giant rainforest seeds, and other souvenirs from my field trips, and the obligatory stuffed caiman suspended from the ceiling. I also have some Indonesian teak bookcases crammed with natural history field guides. Sitting at my Burmese secretaire, I flip through two of these tomes that I have just pulled off the shelf. The first is *A Field Guide to the Birds of West Malaysia and Singapore* by Allen Jeyarajasingam. It once got soaked in a tropical thunderstorm, so some of the magnificent plates by artist Alan Pearson stick together, but fortunately Plate 63 is still in pristine condition. It depicts twelve species of flycatcher, all crisply drawn as pairs of a colorful male flanked by a drab female. The males are easy to identify: each species (sparrow-sized, with a short, strong beak planted in a tuft of springy bristles) has its own unique plumage combination of air force blue, amber, off-white, and sooty black. Most of the females, on the other hand, are a nondescript shade of brown. The second book lying open on my desk is an atlas to the grasshoppers and crickets of the Netherlands. Though the males of many European grasshoppers look quite similar, they sing very different songs to attract females; pages 113 to 115 carry a key to the orthopteran hit parade and tucked in the back of the book is a CD with which the grasshopper enthusiast can learn to recognize species by ear.

If I went on flipping through my field guides, I would come up with dozens of other examples of how the best way to identify otherwise similar species of pheasants, toads, fireflies, chameleons, and so on is by the males' feathered crests, croaking pitch, blinking pattern, skin coloration, etc. Apparently, these male signals ("secondary" sexual characters, as Darwin called them) used to woo females are among the things in nature that evolve the fastest and in widely different directions, yielding the greatest differences between species. Rather than adapting to the environment, these colorful feathers, harmonious voices, and all the other ways in which males attract attention to themselves are constantly adapting to female fancy.

Now I drag another dusty book off my shelf: the third volume of *Die Käfer Mitteleuropas* (The Beetles of Central Europe) by Heinz Freude, Karl Wilhelm Harde, and Gustav Adolf Lohse. Opening it to a random page in Chapter 7, I come upon a series of line drawings that illustrates how to identify the thirty or so species of tiny *Hydraena* water beetles that live in the streams and swamps of Germany and surrounding countries. It doesn't show the beetles themselves, which all look extremely similar. Instead, it shows a plate that looks like it was taken from a catalog of mail-order kitchen appliances: rows and rows of outlandishly shaped water beetle penises, the only way to correctly determine which species a hydraenid belongs to.

The fact that water beetle penis diversity fits in a more general pattern of species differing in their kinds of male adornment should make us wonder whether their genitals might also, like flycatcher plumage and grasshopper song, evolve by Darwinian sexual selection.

This chapter will be devoted to Eberhard's theory that female preferences are indeed what drive the evolution of male genitalia. But first let us ask ourselves why Darwin, after spending so many pages discussing the elaborate feather displays of birds and the immensely varied constellations of horns, antlers, and prongs on the heads and thoraxes of male animals both vertebrate and invertebrate, didn't take the next step and also embrace the priapic parade of animal penises.

Perhaps he was sidetracked by the obvious functionality of the genital organs (in contrast to, say, virtuoso song and pink tufts of down). Remember, the fact that a male simply *needs* a penis (to serve his reproductive needs) is why Darwin categorized genitals as primary rather than secondary sexual characters and, hence, the result of natural rather than sexual selection. But this does not really make sense. A bird also needs a head, but that does not detract from the fact that any colorful attributes *to* its head may be selected by females—it's the "part" versus "property" distinction of

Ghiselin's that we saw in Chapter 1. Similarly, the penis itself may be a functional necessity, but all its ornamental attributes (spines, flanges, ribs, knobs, anything beyond the modest functional needs of a syringe) may be selected sexually.

Perhaps, too, Darwin simply wasn't aware of the magnitude of penis diversity. In his 1985 book, Eberhard thought that this indeed was the explanation for Darwin's silence on the subject: "So strong was Darwin's apparent belief in the powers of sexual selection that if the complexity and variety of genitalic structures had been common knowledge among zoologists of his day, or even, perhaps, if he had studied beetles rather than barnacles, I suspect he would have included genitalia in his listing," he wrote.

But then Darwin *did* study beetles—as a student in Cambridge, he was such a passionate beetle collector that one of his classmates even drew a caricature of Darwin riding an oversized dung beetle while at the same time trying to catch it with a minuscule net. And in his autobiography Darwin spends several pages glorifying those beetle-hunting days. However, he also admits that, aside from pinning and identifying them, he did not study his beetle catches very intensively and performed no dissections, so he may indeed have missed the genital wonders that lay hidden beneath their elytra. Then again, the genital diversity of barnacles, which Darwin did study and dissect with abandon for eight years (because he felt he could not begin his evolutionary work without first having written a decent zoological treatise), had no secrets for him. He was the discoverer of the penis that still counts as the longest in the animal world: the male organ of the burrowing barnacle (*Cryptophialus minutus*), a whopping eight times longer than its body. (This is explained by the fact that barnacles live their lives affixed to a rock or a ship's hull and the only way to fertilize their neighbors, in a wave-swept environment where casting sperm into the water would not work, is by extending long, prehensile penises to seek out mates. A particularly steamy day in the barnacle colony shows dozens of elongated penises snaking their way among the denizens, probing crevices wherever they can. Sometimes there are so many at the same time that it is impossible to determine who is mating with whom.)

No, the reason that Darwin steered clear of all too explicit mentioning of sex and genitals in his books (aside from his publications intended for a specialized audience of barnacle aficionados) was probably Victorian

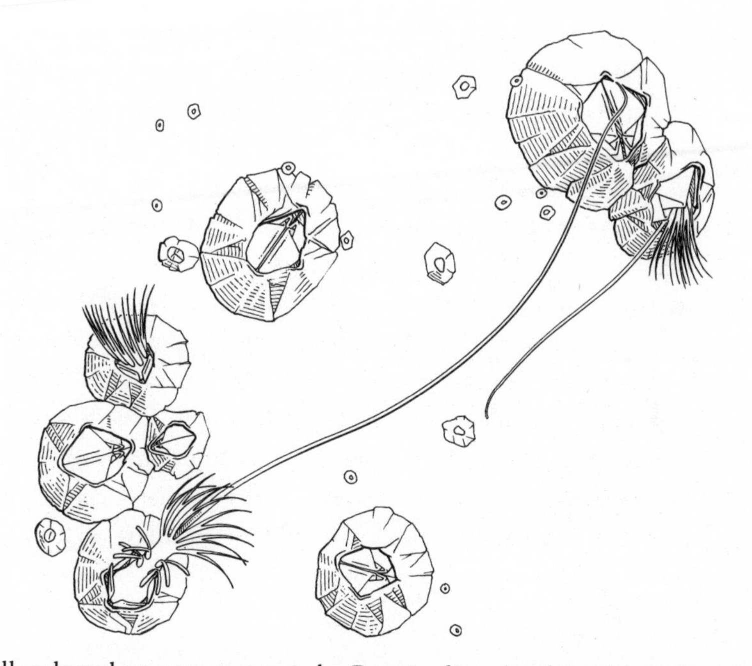

Well endowed on wave-swept rocks. Darwin discovered that barnacles, which cannot approach each other to mate, have the longest penises in the animal kingdom (relative to their body size, that is). Here, the animal on the bottom left is seen penetrating the one on the top right with its snake-like penis, while another one (far right) is just beginning to probe around for a mate.

caution. A popular image of Charles Darwin is often that of a sideburned, sunburnt Victorian explorer in breeches, clambering about the volcanic rocks of the Galápagos Islands in search of evidence for evolution. But the reality is that by the time he celebrated his greatest successes, Darwin did most of his exploring in his head and library, while his daily life was particularly humdrum. Father of seven, he doted on his children and, next to penning down earth-shattering scientific theories, his main concern was to be patriarch of a harmonious upper-middle-class family.

Not surprisingly, these two ambitions sometimes clashed. While Darwin and his wife, Emma, held a long-standing but good-humored difference of opinion over evolution versus divine creation, his struggle with the subject of sex seems to be personified in his eldest daughter, Henrietta ("Etty"). Working on the manuscript for *The Descent of Man, and Selection in Relation to Sex,* Darwin had managed to resist the efforts of his publisher to remove the word "sex" from the title and had rescued some passages (about colorful swollen monkeys' "bottoms," for example) by relegating them to footnotes in abstruse Latin. But he yielded to Etty, who, while proofreading the manuscript, pulled out her red crayon whenever she felt that Dad had strayed beyond the confines of Victorian propriety.

To be fair, although Darwin biographers often portray Etty as a prude with a negative influence on her father's work, her recently published diary shows a much more thoughtful personality, and she was also the one who vehemently fought the rumor that Darwin had converted to Christianity on his deathbed. Still, the story that, later in life, she started a campaign to single-handedly rid the English countryside of the obscenely shaped stinkhorn mushroom (*Phallus impudicus*), because of the bad influence the fungus's appearance would have on maidens, gives one pause. . . .

Darwin and the Cost of Eggs

So the conspicuous absence of genital diversity (except perhaps those monkeys' bottoms) from Darwin's sexual selection writings left the subject up for grabs. And grabbed it was, by Bill Eberhard, albeit somewhat belatedly. But before checking out Eberhard's take on the genitals and sexual selection, let's have a closer look at the theory of sexual selection itself. You probably recall its basic principle from Chapter 1: if a male possesses a heritable "sexy" attribute that gives him the edge over other males in the competition for copulations with females, he will sire more offspring and his heritable attribute will be overrepresented in the next generation; hence, evolution takes place. (And I gave the example of what happens when female birds prefer a fawn-feathered mutant over the regular maroon-backed ones.) Now, you would be completely justified in wondering why the reverse could not also be true: if some females are more attractive to males than others—not unimaginable, right?—wouldn't they also experience sexual selection?

Well, yes and no. It's complicated. So complicated that it has required sixty-five years of scientific papers crammed with mathematical formulae. But don't worry—I'll brief you in four math-free paragraphs. The main character in this evolutionary debate is not Darwin but an English geneticist named Angus Bateman. In 1948, Bateman wrote a paper in the journal *Heredity,* in which he observed: "Darwin took it as a matter of general observation that males were eager to pair with any female, whereas the female, though passive, exerted choice. He was at a loss, however, to explain this sex difference, though it is obviously of great importance." In other words, if it is generally true in nature that males, not females, are the ones with one-track minds, then why would this be so? What causes the sexes to have such different priorities? Although Bateman's paper has caused a lot of controversy, even years after its publication, we have to thank him for spelling out the core issue, namely: the price of eggs.

Bateman's rationale went like this. Imagine a breeding colony of animals—say, a covey of those partridges. Now imagine you could look inside the bodies of those birds. Inside the ovaries of all the transparent females you would see lots of eggs, and inside the testes of all the see-through male partridges you'd discern zillions of sperm. Hold on to that image. In an evolutionary sense, those clumps of sperm and eggs floating in midair are the raw material that is going to make up the colony's next generation, and the transparent bodies of the birds that carry them are the vehicles that are going to get them there.

As you gaze at those ghostly creatures with their valuable cargo, you will notice that the eggs are relatively big and few whereas the sperm are small and numerous (we know why this is—remember the organelle wars of Chapter 1?). Since for each new partridge only a single egg and a single sperm need to unite, there is a surplus of sperm. In fact, the sperm from a single male would probably be more than enough to fertilize all the eggs of all the females in the colony for many years to come. This means that eggs, not sperm, are the sought-after commodity and, evolutionarily speaking, the sperm should make their males do their best to ferry them to the maximum number of eggs. This basic inequality leads to an asymmetry in sexual selection: males benefit from more copulations, whereas females don't—or, at least, not as much. So anything that makes a male more attractive to females will be sexually selected. Females, on the other hand, are always desirable to males, as long as they carry eggs that still need to be fertilized.

This, said Bateman, leads to "undiscriminating eagerness in males and discriminating passivity in females." And he proved his point with a series of experiments with banana flies in milk bottles. He placed males and females in the bottles, provided plenty of food, and counted the number of flies that were born to a female if she had mated once, twice, or more often, and also counted how the number of offspring a male sired depended on the number of times *he* got to mate. The results were unequivocal: females always produced a few dozen larvae, no matter how often they mated. For males, on the other hand, their success in fatherhood depended entirely on their sexual prowess: the more copulations, the more children. As a consequence, the numbers of babies, while more or less the same for all females, varied from a measly ten or so to several hundred among the males.

Bateman's principle—"sperm are cheap, eggs are expensive, so females are choosy and males are wanton"—has since then become a central tenet of sexual selection theory. But it has also been criticized. Not only have later experimentalists found flaws in Bateman's fly experiments, but so many exceptions have been discovered of big and expensive (not small and cheap) sperm, of cheap (not expensive) eggs, of choosy (not promiscuous) males, and of wanton (not finicky) females that some scientists argue the principle is violated so often that it should be binned. I think that would be too drastic. Granted, the world is more complicated than the one created by Bateman in his milk bottles. Yet the basic tenet of Bateman's principle (and, indeed, perhaps we'd better tone it down to "Bateman's Rule of Thumb") still stands: the investment per child is *usually* less for males than for females and it therefore *normally* pays for females to be choosier than males.

So much for Bateman. Now back to Darwin, who simply took it for granted that males would be inclined to chase females rather than the other way around. Still, he recognized two different ways in which the chasing could be done: through bullying other males or through wooing females. As for bullying, he said, "The strongest and most vigorous males, or those provided with the best weapons, have prevailed. . . . Through repeated deadly contests, a slight degree of variability . . . would suffice for the work of sexual selection." Anyone who has ever watched stags (or stag beetles) battle over a female would not contest the sense in Darwin's words.

Wooing has been more problematic. I happen to be writing this paragraph on a balcony in Borneo and the star fruit tree I am overlooking is the territory of a lively crimson sunbird male who spends much of his time sipping nectar from the purple flowers. However, whenever a (drab greenish gray) female visits his tree, he temporarily abandons sucking from his favorite flowers and instead starts sucking up to her: hopping from twig to twig, quivering his wings, and seemingly making sure she notices how nicely the sun bounces off his brilliant red, yellow, and metallic purple plumage. Sunbirds do not use their colorful attire to bully rivals; they are dandies that use it to dazzle the female senses. As Darwin wrote, "The females are most excited by . . . the more ornamented males, or those which are the best songsters, or play the best antics. . . . In the same manner as man can give beauty to his male poultry, so it appears that in a state of nature female birds, by having long selected the more attractive males, have added to their beauty."

The reason that this particular version of sexual selection did not go down too well in Darwin's day was that, in those male-biased Victorian times, women—and, by extension, all females—were supposed to be docile and not desirous or capable of making any choices of their own, sexual or otherwise. Trying to preempt such criticism, Darwin wrote, "No doubt this implies powers of discrimination and taste on the part of the female which will at first appear extremely improbable," and then went on to show that female animals were, in fact, quite capable of such delicate choices among ornamented males that, to the human eye, all looked more or less the same. Still, it took until the sexual revolution of the 1960s before female choice became a fashionable topic of zoological research, and biologists started to perform experiments that eventually proved Darwin's dandy theory correct: female animals—be they primates, birds, insects, or spiders—do indeed often prefer to mate with males that look particularly pretty, are sweet voiced, or carry themselves in a "beautiful" manner.

What remains a bit of a mystery, even today, is *why* females would prefer a male with, say, an extra-long red crest on its head. What good does it do her? For a while, debate about this was carried on from two entrenched camps. There were the "Fisherians," taking their nom de guerre from the legendary English evolution theorist Ronald Fisher, who thought that females simply were, as Darwin said, "most excited" by the more elaborately adorned males and chose on the basis of arousal per se. Opposite the

Fisherians stood the "Good-Geners," who found it hard to accept that females would make such arbitrary choices as to prefer males with long red crests simply "because they look smart." They suspected that females used these sexual ornaments to gauge males' "genetic quality," so that a male with a particularly striking ornament would also be particularly strong or healthy—and be able to supply the offspring with those good genes in his sperm.

Only in the past ten years or so have biologists begun to realize that the difference of opinion between the Fisherians and the Good-Geners is only an illusion. It is immaterial what kind of benefit a female's offspring get from her mate choice. If her sons are sexy because her mate was sexy, they will benefit by being as successful suitors as their father. And if her offspring are extra strong and healthy because their father's attractiveness signaled his superior disease-fighting and survival genes, they will also benefit. So although in different species different kinds of benefits are gained by mating with an attractive male, the genes that females go for are always "good."

Martine Maan, then working at the University of Texas, now at the University of Groningen, revealed that in the poison arrow frog *Dendrobates pumilio* from the Bocas del Toro archipelago in Panama, females benefit from both variants of goodness. These tiny frogs accumulate deadly toxins in their skins (they do not make these toxins themselves, but steal them from the ants they eat) to prevent themselves from being eaten by birds and other animals. To warn predators of their inedibility, they have evolved striking warning colorations: orange with black spots, red with blue legs and arms, or yellowish green with black blotches, for example—throughout the archipelago, frogs on different islands have different color patterns. Maan and her colleagues discovered that the more brightly colored frogs were also the most poisonous (which a battalion of unfortunate lab mice proved for them). "Thus, potential predators can tell from the colors of the frogs how toxic they are," explains Maan. They then gave females a choice between two males, one of which seemed brighter than the other. (I say "seemed" because the researchers—to make sure they were really testing only the effect of brightness—used two equally bright males, but put one of them literally in the spotlight.) As it turned out, the females mostly preferred the males that appeared the brightest to them. So, in the real world, even if the females simply *liked* the more brilliant males in a Fisherian sense, the effect was that their offspring would inherit the good genes that gave them better protection against predators.

In the bird world, such marriage of love *and* convenience is also commonplace. Peachicks fathered by the most attractive males with the longest trains grow faster and survive better, while barn swallow males with the longest tail feathers possess the best genetic tools to fight off infection. But we can't assume all male ornaments that win over females signify good health and long life. Some sexual selection by female choice is just about being irresistible, as we shall see shortly.

Feathers and Phalluses

If anyone knows the value of serendipity, it must be behavioral biologist Nancy Burley of the University of California, Irvine, who can trace much of her career to a botched experiment back in the late 1970s. At the time, Burley, then a young postdoc at the University of Illinois, Urbana, was researching breeding behavior in Australian zebra finches. These birds nest communally, so to be able to recognize individual birds in the large aviary she was going to house them in, she fitted each of her forty experimental birds with a unique combination of seven colored rings on their legs. Then something unexpected happened: five months into her project, some males still had not paired off, and those were invariably the ones without any red or pink bands on their legs. When she took some of those frustrated bachelors and added two red bands to their already overloaded legs, they instantly paired off (and as soon as they became aware of their increased desirability, they began philandering and generally behaving as if they were God's gift to female zebra finches). Clearly, the plastic rings on males' legs could boost or break the males' sex appeal: red and pink rings, and the females would be flocking to them; green and blue, and they would turn away in disgust.

Now, a lesser researcher might have cursed such an unexpected turn of events, rounded up her finches, and started all over again with neutral-colored bands—after all, an experimenter should not be influencing the behavior of her study subjects. Not so Nancy Burley. She realized that she was actually on to something very interesting. After all, zebra finches normally don't have brightly colored legs, so the females in her aviary were choosing their mates based on an aesthetic preference that they normally never got to exercise. Intrigued, Burley began experimenting with other artificial ornaments for her male finches. She dyed loose feathers to produce white, red, or green crests, which were then fixed upright on top of the males' heads with a crown of beads and a generous dollop of glue. And even though crests do not naturally occur in zebra finches or in any of their relatives, the new headgear once again had females swooning—but, as if to underline the whimsicality of "a woman's reason," only the white version: the red and green ones left them cold.

The surprising finch aesthetics that Burley unveiled are grist to the mill of a version of sexual selection theory known as sensory drive. The idea is that, obviously, animals must choose their sex partners on the basis of what they can gauge of them with their senses. In every species, the working range and resolution of those senses have been optimized for the species' habitat and way of life, albeit limited by what is "technically" possible

A woman's reason. Female zebra finches swoon over males with white-crested crowns—even when those crests were glued on by a researcher.

in the species in question. Take vision. The zebra finches of Nancy Burley munch seeds and fruit and have multicolor vision to find and distinguish these in the surrounding foliage. Their retinas carry four different kinds of color pigments, with high sensitivities in ultraviolet, blue, green, and red. These sit in cone cells, each of which is wired via neurons to the zebra finch brain. Rather than such four-channel vision, humans have only three color pigments (for blue, green, and red), most monkeys two, and the owl monkey eye can create only monochromatic images; mantis shrimp, on the other hand, have at least seven, including three ultraviolet ones. Other animals forgo vision entirely and instead live in a world of sound or scent, hearing, touch, or more exotic senses, like the platypus's electroreception.

In other words, we can view a female animal (for this particular purpose) as a central processing unit with a complex series of input channels. Each of those input channels has different sensitivities, and stimulation within those sensitive ranges will result in a physiological response. One of the peak sensitivities of the zebra finch eye, for example, is an orangey kind of red, so a male with an ornament (either real or artificial) of the right sort of red *will* stimulate the neurons in her eye, whether she likes it or not. Or rather, sensory drive adepts would argue that stimulation *equals* liking. Whatever a male can produce during courtship that strikes her eye, caresses her ear, or just generally dazzles her senses will be registered, noted, and, by default, liked.

It is at this point that the phallus lumbers onto the scene. Bill Eberhard's revolutionary idea was that the main role of the penis on the stage of reproduction is not just sperm transfer (that, too), but rather courtship, played out in a theater of sensory drive. Females, Eberhard claimed, employ their senses to exert their preferences over the available phalluses just as they do for other available male ornaments. Occasionally, females do this on sight. In 2013, Brian Mautz and his colleagues used animation software to project life-sized computer-generated nude men on a screen and asked women to rate their attractiveness. These volunteers clearly turned out to have a preference for virtual male models with large penises, especially in combination with tallness or masculine hip-to-shoulder proportions. And in some other vertebrate animals like lizards, mosquito fish, and primates, males actually show off their penises for the femalefolk. For example, in a 1963 paper in *Animal Behaviour*, researchers Detlev Ploog and Paul

MacLean described "display of penile erection" in squirrel monkeys this way: "[A] male approaches a female or another male head on, places one or both hands on its back and thrusts the erect penis towards the face. In doing so it 'thigh-spreads' [and] the gaze is directed away from the recipient." (And they illustrate the behavior with a touching little pen drawing.)

Primate exhibitionism aside, the evolution of male genitalia through sensory drive mostly proceeds via the female's tactile senses, thinks Eberhard. And his reasons are quite persuasive. To begin with, there's the penis's complexity. We have already seen that the organ possesses all kinds of bits and pieces that seem unnecessary for the mere task of squirting sperm into a female. In fact, if that were really its sole function, says Eberhard, then the penis of many an animal would be akin to a Rube Goldberg machine. (The hilarious inventions of cartoonist Rube Goldberg are composed of immense series of interconnected parts to perform an incredibly simple function—like the nineteen-part pencil sharpener comprising, among other components, a kite, an old shoe, and a live woodpecker.) But nature has no sense of humor, so the constituents of the penis must serve some more meaningful goal. And, simply put, that goal seems to be to tickle the female in as many ways as possible. To begin with a rather convincing example, published by Eberhard himself in 2009: a crane fly with a vibrator on its penis.

On a cool December morning in 2006, Eberhard, who, next to his position at the Smithsonian Tropical Research Institute in Panama, also works at the University of Costa Rica, found a pair of crane flies (probably a new species of the subgenus *Bellardina*) having it off on the wall of a house in San Antonio de Escazú. As is usual in these insects, the male was dangling underneath the female. Always eager to expand his catalog of animal intercourse, Eberhard coaxed the female onto a twig and rushed the pair to the lab, where he placed them under a microscope and videotaped the whole affair—not anticipating the surprise they had in store for him. In the November 2009 issue of the *International Journal of Tropical Biology,* he and crane fly specialist Jon Gelhaus tell their story of sex, flies, and videotape.

To begin with, the meshing of the pair's genitalia was rather complex. The male crane fly's aedeagus (to use Jeannel's term) is a tightly packed bunch of tubes and plates. Several of these form a cylinder through which, during mating, two appendages of the female's abdomen, so-called cerci, are inserted. Thus configured, the pair that Eberhard had arrested first remained motionless for the better part of fifteen minutes. Then, suddenly, a subtle, repetitive movement began to take place, which, under the microscope, revealed itself as the first confirmed case of "copulatory stridulation" by a male's genitalia. On the outside of the aedeagus, the male carries a pair of hairy, claw-like things called outer gonostyli. These it moved rhythmically to and fro along the outside of that cylindrical part of his aedeagus encasing the female's cerci. On closer inspection, the part of the cylinder that the claw scraped across turned out to be covered in small, parallel ridges, perpendicular to the direction of stroking. In other words: a miniature washboard, scraped, in repeated two-second bursts fifteen to thirty times per second, to produce a vibration with a pitch slightly below middle C. The female, her cerci firmly gripped in the washboard itself, must have felt this hum reverberate through her entire genital region.

This is probably not an isolated case. Many other species of crane flies, as Eberhard and Gelhaus discovered, have such a file-and-scraper system on the male genitals. Sometimes the file has the same frottoir-like appearance as in the male that Eberhard filmed. In other species, the ridges are replaced by bumps or puckers, probably producing a whole different sensation. And similar devices are found on the male genitals of several families of moths. For some of these, it was already known that the males use them to sing at an ultrasonic pitch to communicate with females from a distance (the sensory drive here being that moths possess ears that can detect the ultrasonic clicks of the bats that chase them). But Eberhard was the first to actually see genital stridulation *during* mating. In fact, after he separated the couple and "collected them" (the entomologist's euphemism for dunking an insect in a vial of alcohol), he noted drily, "The male outer gonostyli continued to execute the scraping movements in slow motion as the male expired."

And the vibrator-penis of the crane fly is just one highlight of the animal world's multipage sex-aid catalog. Remember parameres? Those whip-like or drumstick-like appendages of many a beetle penis that Jeannel used to classify his cave beetles? Well, whip and drum is exactly what they do. The penises of beetles, flies, butterflies, moths, and many other insects often carry such accessories (although they are called by different names in different types of insects), and they are used to drum, tap, slap, or stroke

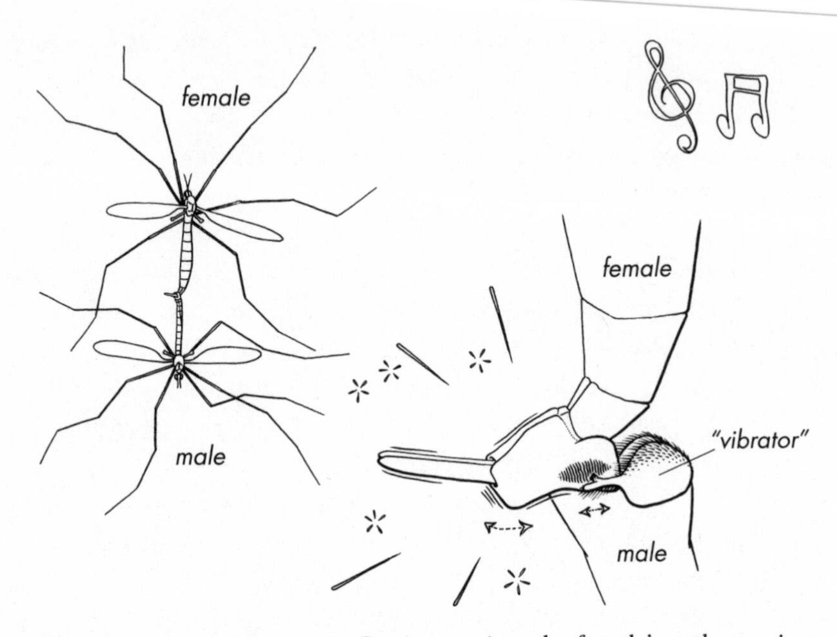

A crane fly with a singing penis. During mating, the female's nether regions are stuck through a sleeve on the male's genitals. These have a washboard-like contraption that produces a vibration with a pitch just below middle C.

the outside of a female's genital region during sex while the rest of the aedeagus is working on her inside. Susanne Düngelhoef of the University of Bonn in Germany studied the use of parameres in leaf beetles and noted that during mating a female often would reposition a male's parameres with one of her hind legs. Not forcefully, as if to kick him away, but more gently ("More to the left—yes, that's it"). The parameres are particularly large and flat in some ladybird beetles, such as the common American spotless ladybird *Cycloneda sanguinea.* By inspecting a mating pair up close it is possible to see the hairy parameres literally slap the female's genitals like two miniature Ping-Pong paddles. As all this happens out of view of the female, sensory drive adepts think that it is the triggering of nerve endings in her hindquarters that makes the female pay attention to what the male is "telling" her.

Now, you could argue that since the crane fly vibrator and ladybird parameres stay outside of the female, they don't really qualify as what Eberhard calls "internal courtship devices." But parts of the male genitalia that *are* inserted deep into the female also seem to have a job to titillate her. For example, the penises of many insects but also some mammals have a whip-like extension at the tip. In hoofed mammals, for instance, the urethra leaves the penis on the left-hand side and continues inside a thin, worm-shaped appendage (in rams, this appendage adds another 4 centimeters—1.5 inch—to the length of the penis). In bulls, it has been seen that this so-called urethral process flips forward inside the female upon ejaculation, probably triggering some kind of sensation in the cow. And then there are all the other outlandish adornments such as spines and knobs on the penises of snakes, monkeys, and cats, and multiple sideways-pointing flanges on the rhinoceros phallus (which the men of some indigenous peoples of Borneo emulate by piercing their own penises with a crosswise metal rod, the so-called *palang*). Not to forget the seemingly rigid but apparently flexible plates of wasp penises, which Eberhard's wife, the famed entomologist and evolutionary biologist Mary Jane West-Eberhard, has described as capable of "the most fluid and subtly modulated movements I have ever observed in wasps."

Speaking of movements: the genital thrusting that to us seems almost synonymous with copulation itself speaks in favor of Eberhard's hypothesis. After all, familiar as the old in-out may seem, there is no obvious reason why genitals should be moved rhythmically at all for sperm transfer: animals eject all kinds of liquid—venom, urine, feces, slime—without any need for rhythmic movement in the ejecting organ in question. Yet in three-quarters of all species of insects, spiders, and mammals—and also in snakes, millipedes, and roundworms—definite humping goes on. And not only before ejaculation: many species continue to thrust for a long time *after* the male has ejaculated (in the case of the thick-tailed bush baby, for up to four and a half hours). Even in some insects that mate without any outward motion visible, the penis often moves to and fro or throbs inside the female, which Eberhard calls "cryptic thrusting." Rather than to make the release of sperm possible, it seems more plausible that these movements are the most efficient way for a male to use his penis as an internal courtship device and make the female sense whatever knobs, ribs, spines, and *palangs* it is provided with. A male that is able to stimulate the receptors of nerves in his mate's inner sanctum particularly strongly will leave a lasting impression, increasing his chances to be a preferred suitor.

It is telling that songbirds, which do not have penises and mate simply by a fleeting "cloacal kiss" (the pressing against each other of the male and female genital openings), copulate for just a few seconds without any rhythmic movements. Except, that is, the buffalo weaver, the only songbird that does have a kind of penis—a rod 1.5 centimeters (0.6 inch) long in front of its cloaca—which it rubs against the female's cloaca for up to half an hour before it, with quivering wings and clenched feet, climaxes.

Of course, for internal courtship to work, the female needs to have the appropriate wiring in *her* genitalia to register these subtle shapes and movements. Sadly, the study of vaginal sense organs is still in its infancy. But where people have looked, they mostly have found evidence for very delicate sensitivities in the female nether regions. In damselflies, for example, the inner walls of the vagina carry a plate on either side, each of which is covered with dozens of separate sense organs. And cockroaches belie their reputation as crude vermin in the intricate arrangement of so-called campaniform sensilla in the female genitals—tiny organs that can register subtle buckling and bending of the chitin that her vulva is supported by. But of course the best-known (though still, as we shall see, understudied) female genitalia are those of humans. A woman's vaginal walls are uniformly innervated with two kinds of nerves (large and small) all along their length, and the furrow between the foreskin and the glans of the clitoris is even more richly endowed with a variety of microscopic organs that sense fine touch, pressure, and vibration and carry such exotic names as Meissner's corpuscles, Pacinian corpuscles, Krause's end-bulbs, Ruffini's corpuscles, and mucocutaneous corpuscles. All prone to be exploited sensorially by a moving, complicatedly shaped male organ.

So there we have it. Assuming that the shape, size, and adornment of a phallus are to a large extent heritable (and where this has been studied, they tend to be), sexual selection by female choice will be a powerful force by which the evolution of genitalia is pushed around. But hang on: throughout this section you may have sensed a nagging feeling that something doesn't quite add up. Because isn't there a crucial difference between courtship and copulation in the sense that during regular courtship rituals a female can still reject a male, whereas once she has allowed a male to insert his penis into her, isn't it sort of too late to think, "Hmm, maybe I won't let this particular male fertilize my eggs"? Well, prepare to have another of your certainties about sex squashed.

Love's Labor Lost

Scientific research papers are usually not particularly riveting reading material. The Materials and Methods section is often the least appealing, easily skipped in pursuit of more enticing bits. But it is in the "M&M" (as scientists call it) that, hidden under dull jargon and cloaked in the passive voice, the jewels of original scientific research lie. Condensed beyond recognition into terse, matter-of-fact sentences are days, months, sometimes years of hard, enjoyable, or exhilarating (but often also mind-numbingly dull) labor, moments of despair and jubilation, false starts, near give-ups, luminous ideas, and innovative inventiveness. And not infrequently a generous helping of courage.

The M&M sections of the two articles that Robin Baker and Mark Bellis published in the November 1993 issue of the journal *Animal Behaviour* describe some of the most courageous biological experiments of all time. Not courageous in the way that clambering up tropical cliff faces in search of rare animals or plants is courageous—Baker and Bellis never left their offices. Instead, courageous because they dared ask their closest colleagues and students to become sexual guinea pigs for them. In dry, to-the-point tone, the M&M section describes what they did: "[W]hole ejaculates were collected in condoms during copulation or masturbation. Subjects were provided with a 'kit' containing instructions and all necessary equipment, including a mixture of lubricated (non-spermicidal) and non-lubricated condoms. Ejaculate collection and fixation and the counting of sperm followed the double-blind protocol." It states furthermore: "[C]ounts of sperm [were made] of 'flowbacks' (the mixture of seminal fluids, sperm, female secretions and female tissue that flows back out of the vagina after copulation)."

Although Baker went on to write the partly pornographic, partly popular science book *Sperm Wars* and was heavily criticized by some of his colleagues for blurring the distinction between scientific fact and science fiction, those two 1993 papers by Baker and Bellis still stand. They are, so far, the only experiments ever performed on "flowback" in humans.

What is "flowback"? Rather than just the unavoidable consequence of the messiness of copulation, flowback—or, more crudely (but probably more accurately), "sperm dumping"—is now seen as one of many different ways in which females can actively control which male gets to fertilize her eggs, even after having been inseminated. Throughout the animal kingdom, females are seen to eject anything from droplets to gushes of semen immediately after copulating with a male, and humans are no different.

Baker and Bellis somehow persuaded their colleagues and students at the University of Manchester to share with them—quite literally—something very personal. For several months, thirty-five heterosexual Mancunian couples (consisting of thirty-three men and thirty-three women—there was some partner swapping going on) obediently dropped into Baker's mailbox envelopes bearing the liquid testimony of their sex lives: properly labeled and tied-up condoms with ejaculate resulting from either masturbation or protected vaginal sex, as well as condoms with collected flowbacks if sex had been unprotected (women were advised to collect these by squatting over a beaker and coughing). The couples also dutifully kept logbooks of their sexual activities and specifically of the woman's orgasms. In the end, Baker and Bellis were the proud owners of a freezer filled with unique samples: 67 masturbation ejaculates, 84 copulation ejaculates, and 127 flowbacks. It was time to pull out their microscopes and start counting and calculating.

For each sample, they used standard microscopic methods to determine the sperm content. They first had a look at the two types of ejaculates—those deriving from masturbation and from intercourse. Based on the sperm counts and the information written up in the logbooks, as well as other personal details, Baker and Bellis were able to work out, for each individual man, a formula that predicted quite accurately how many sperm there would be in his ejaculate based on how long ago his previous masturbation and/or previous copulation had been, how much time he and his partner had spent together since they had previously had sex, and, rather surprisingly, the weight of the woman. Then they turned to the flowback samples.

They also counted the number of sperm in the flowbacks. Of course, there was no way they could directly determine what proportion of semen had been dumped by the woman, because the remainder had been left behind inside her vagina and uterus and had not made it to Baker and Bellis's freezer. But they had their magic formula, and with this they could calculate quite precisely for every flowback how many sperm must originally have been in the man's ejaculate. The results were so intriguing that when the duo presented its work at a congress in Princeton, the auditorium was packed, "while speakers in a parallel session were talking virtually to themselves," as one commenter recalls.

First of all, some semen was dumped by any woman after almost every copulation. And although *on average* about 35 percent of the inseminated sperm was flushed out again in the flowback, there was a lot of variation to this number. Sometimes the flowback contained almost zero sperm, so the woman had retained the entire ejaculate inside her reproductive system. At other times (about 12 percent of all the flowbacks) she managed to return virtually all the sperm her mate had just donated to her. In other words, in theory, sperm dumping could provide a woman with a way to make or break a man's reproductive success. But did she? And if so, how? This is where the orgasm comes in.

Baker and Bellis found that the proportion of sperm the woman expelled was determined by her orgasm. If she did not climax or if her climax was more than a minute before the man's ejaculation, she retained little of his sperm. If, however, she had an orgasm during or after the man's ejaculation, she retained a lot. In other words, women can "use" their orgasms as one way to manipulate the likelihood of a particular man fertilizing an egg of hers. (Even in humans, the word "use" should not be construed to imply that this is a conscious decision; rather, it is the evolved complexity of our physiology that makes this happen involuntarily.) We will come back to women's orgasms later (yes, this book will provide you with multiple orgasms!), but for the moment let's leave our own species behind and turn to sperm-dumping females in somewhat less prominent species. As we shall see, human females share the habit with many, many animal species.

Nobody knows whether females of *Silhouettella loricatula* experience orgasm, but they certainly do sperm dumping. These miniature "goblin spiders" live in soil and in the leaf litter underneath carob trees, which is where Swiss arachnologist Matthias Burger found them. When he took some males and females to his lab at the Natural History Museum of Bern, he noticed how the females would release a tiny bag of sperm from their genitalia during the copulation with a male. If you recall from Chapter 1, spider males mate by first loading their pedipalps or "sex legs" with sperm, and then working these from below into a female's genital opening. This is not, by the way, a simple hit-and-run affair. In fact, copulation can be a lengthy process during which the male makes all manner of complex prying and twisting movements (internal courtship?) with his pedipalps.

Burger discovered that, unlike in human females, the sperm dumped by his spider ladies was not that of the current male, but of her previous mate. To find this out, Burger had to do two things. First of all, he had to flash-freeze copulating pairs by pouring some liquid nitrogen over them—about minus 200 degrees centigrade (−328°F); talk about a cold shower!—and then performing painstaking dissections of the genitalia "frozen in action." (Awe is in order: these spiders are about 1.5 millimeters long—that's 0.07 inch!)

What Burger discovered was a surprisingly sophisticated sperm-processing mechanism in the female. Whenever a male's pedipalps would gain access to her genitalia and squirt in a droplet of sperm, they would not get any farther than a vestibule, the so-called receptaculum. Beyond, closed off by a solid valve and unreachable to his pedipalps, lay the uterus, where eggs could be fertilized. Burger found that lining the receptaculum was a sheet of glands with which the female would secrete a capsule around the male's sperm. At its tip, this capsule was extended into a narrow tube, which was clamped between plates leading to the uterus. In other words, the female goblin spider packages sperm from different males separately and can, at least in principle, "decide" (note the quotation marks) whether she will use the sperm to fertilize her eggs or will instead dump the whole bag when a better spider (one that wields his pedipalps even more elegantly, for example) comes along.

Silhouettella loricatula and *Homo sapiens* are just two of the many species known to practice sperm dumping. It has been studied in other spiders (including the well-known cellar spider *Pholcus phalangoides*, which lives in houses all over the world and shakes its web vigorously when disturbed), lots of insects, and other mammals. Female Grevy's zebras, for example, release a large amount of semen after mating, and the zoologist Joshua Ginsberg measured (quickly, before they would be soaked up by the dry soil of the Kenyan plains) diameter and depth of these puddles of semen. He estimated that they amounted to an average of 0.3 liter. Rebecca Dean and colleagues performed similar experiments in a Swedish chicken coop, where they videotaped more than a thousand copulations and, after analysis of the footage from two video cameras that had been trained on the females' cloacas, determined that females dumped more sperm if they had been mounted by cockerels lower in the pecking order.

Even such humble creatures as nematode worms can make sophisticated "choices" about keeping or dumping sperm. *Caenorhabditis elegans* is a microscopic soil-dwelling roundworm that has been the workhorse of embryologists since the 1960s. Given its simple millimeter-long (0.04-inch) body made up of exactly 959 cells, you wouldn't expect the many intricacies of its mating behavior that David Barker lovingly describes in a 1994 article. When a male meets a female, Barker writes, he extends his so-called bursa, a fold-like extension of skin around his genitalia, wraps it around her, and inches up and down the female until he has located her vulva. Worms being what they are, there are not many orientation points along her body, so sometimes the male searches up and down her flank until, puzzled, he decides to try the other side and there finds what he is looking for.

Having located the vulva, he then inserts his spicules. These are essentially the nematode worm's penises. Unlike the rest of a worm's squishy body, spicules are tough and rigid, of complex and tortuous form (as we now have come to expect of penises). They often come in pairs, and are used to open the female's vulva, and also, again, for something that resembles internal courtship: a repetitive thrusting motion preceding the actual two-minute-long copulation. (Given the worms' four-day generation time, this is not as fleeting as it may sound.) These small nematodes are helpfully transparent, so Barker could observe exactly what happened to the sperm inside the worms' bodies once copulation had started. Ten seconds after he had stuck his spicules into the female's vulva, the male's sperm began to collect in the tubules leading from his testis to his spicules and was then, in small bursts, pumped into the female's vagina. Toward the end of the two minutes, the amounts of sperm transferred became smaller and smaller and eventually the male began to withdraw his spicules. At that point (and it is hard not to imagine Barker staring wide-eyed into his microscope eyepiece and uttering a muffled expletive), the female would sometimes return the whole amount of sperm just received in a manner suggestive of blatant refusal. Barker writes: "The vulva would open and the entire mass of seminal fluid would appear to be blown out of the uterus under pressure. This usually resulted in the spicules of the male also being blown out of the vulva."

The widespread fact that insemination, as for these male worms, sometimes quite literally backfires leads to what Eberhard has called "a sad conclusion from a male's perspective," namely that copulation does not necessarily lead to insemination and fertilization. Just like putting a coin in a slot machine does not automatically lead to a jackpot, inserting a penis into a vulva is just the start of a hurdles race in which the risk of sperm dumping is just the first obstacle. As Eberhard has written, "[F]emales, because fertilization takes place within their bodies, generally have the last say in reproduction and can exercise 'cryptic female choice.'"

As Bateman has shown us, a unit of sperm is, to a male, relatively cheap to donate and any copulation is worth the shot, just like sticking a coin in a slot machine and not winning anything is an acceptable loss to any gambler. For females, the balance is different; she has a limited number of eggs to lay, so she has to hand out those jackpots prudently—either only to the males with the best genes or by spreading them evenly among males so as to create some genetic diversity among her clutch. It is, therefore, in the female's interest to retain control over the series of hurdles between his ejaculation and her conception and turn each of those hurdles into points of decision in her favor. That's why sexual selection does not stop with copulation. After a female has used her entire sensory apparatus to gauge a male's assets to decide whether she allows him to mount her, there is a whole range of sexual selection opportunities after copulation has begun, and sperm dumping is one of them. Besides sperm dumping, female animals (including humans) employ a broad range of additional filters to select from among the males whose penises their vaginas encounter. And in the next chapter, we learn a few more of these filtering tricks.

Chapter 4

Fifty Ways to Peeve Your Lover

Well, perhaps not fifty ways, but at least twenty-three, says Bill Eberhard in his 1996 book *Female Control: Sexual Selection by Cryptic Female Choice.* This book, a successor to his *Sexual Selection and Animal Genitalia* of a decade earlier, served a single purpose: to point out that, counterintuitively, when a female mates with a male, this should not be taken to mean that she has yielded to his desires to sire her offspring. Copulation is not "the criterion for final acceptance in female choice," Eberhard writes. Rather, it is just one stage somewhere in the middle of a whole concatenation of hurdles put in place by the female. These hurdles start way before copulation, where they have been the territory of traditional studies of sexual selection. But—and this, certainly back in the 1990s, was not yet fully appreciated by biologists—they also continue long after copulation has begun. Copulation does not necessarily lead to insemination, insemination may not mean fertilization, and fertilization is not the same as reproduction. Each of these steps is, in Eberhard's opinion, controlled by the female in a set of strategies he calls "cryptic female choice"—cryptic because these decisions are taken deep inside the female's body, out of view of her mate and human observers alike.

Ignoring this internal extension of mate choice, says Eberhard, is a serious error on the part of biologists. Females are not "passive offspring-generating machines" but, instead, very sophisticated products of evolutionary selection that have been optimized for serving their own and their offspring's interests. And such optimization applies to reproductive decisions taken before as well as after a male has begun to insert his penis into her. In the stage before copulation, decisions are taken by the female on the basis of a broad range of male signals (his color and size, pheromones, courtship dances and song, to name but a few) perceived with a broad range of her senses (vision, hearing, smell). But after copulation has begun, the male signals as well as the female senses and decision-making machinery involved shift to those concentrated in and on both partners' genitals. And although genitals are just a small part of the body, this does not mean that the evaluation of the signals and the ensuing decisions are any less important or effective.

Sperm dumping is one effective (though perhaps not particularly subtle) way by which females can choose from among the males that mate with her. In this chapter, we will explore the many other strategies that females have in their arsenal to thwart the reproductive intentions of certain suitors.

One particularly cunning strategy is the kind of two-stage coitus that is practiced by certain rodents. Neotominae are a group of 130 species of native American mice that includes deer mice and pack rats, and we would probably still be in the dark about the sexual preferences of these small mammals if it weren't for the tireless efforts of one researcher, Donald Dewsbury of the University of Florida. Starting in the late 1960s, Dewsbury published almost thirty articles all entitled "Copulatory Behavior in the . . ." (fill in your species of neotomine mouse). Each would give detailed observations for the mouse in question on duration, sequence, and frequency of discrete elements in rodent sex, such as "pursuit-mount," "intromission thrust frequency," "ejaculation latency," "genital grooming," and (my favorite) "lying down." One can only imagine the dull fatigue that may have overcome a journal editor upon receiving yet another brown envelope with a mouse manuscript by Dr. Dewsbury, which is presumably why he targeted such a broad range of different journals, like *Animal Behaviour, American Zoologist,* and *Journal of Mammalogy.* Still, taken together, Dewsbury's articles form an unrivaled sort of Kinsey Reports for rodents, and they show some noteworthy patterns—patterns that may reveal junctures during copulation at which a female mouse can exert that cryptic choice of hers.

In most neotomines, the genitalia lock together during mating because the mace-like penis inflates and cannot be extracted, meaning sometimes the male drags the female around long after ejaculation, until finally his penis becomes flaccid. In some species, such locking is replaced by a more female-friendly style of copulation. A male California mouse (*Peromyscus californicus*) will mount a female several times in succession. Each time, he will grace her with a series of "deep pelvic thrusting" and then dismount, only to remount a second time and do the same, and sometimes a third, each time without ejaculating. Only on the final series of thrusts would he climax, easily recognizable "by the presence, after several deep thrusts, of a series of spasmodic muscle contractions with rapid shallow intra-vaginal thrusting, and a long period in which the male remains immobile, clutching the female." Awww.

Such bouts of "dry sex" preceding "wet sex" might be a particularly suitable way for females to select males. Using cues from his genitalia only—or, as Eberhard calls it, "copulatory courtship"—a female could, after the dry phase, decide to allow or disallow the crucial wet phase. And copulation in two or more stages is surprisingly common in a wide variety of animals. Tiger beetles do it, and so do certain moths, wasps, stick insects, jewel beetles, mites, rats, and some spiders. The discovery, back in the 1960s, of dry copulation in the sheet web spiders (a large family of small spiders, many of which build those small sheet- or hammock-like webs that collect early-morning dewdrops in tall grass and shrubs) was a bit of a surprise, at least to spider specialists. As you may recall from Chapters 1 and 3, the received wisdom about spider sex is that a male will build a small web, squirt some sperm into it from his genital opening, then suck up this sperm in his often bizarrely shaped pedipalps and wander off in search of a female to inseminate. This is how many species of spiders go about mating, so naturally Dutch spider specialist Peter van Helsdingen expected to find it when, in 1962, he began his undergraduate project on the mating behavior of the money spider *Lepthyphantes leprosus*, a European species from the sheet web spider family that dwells in caves and cellars.

Van Helsdingen obtained the spiders from the artificial caves left after centuries of marl excavation in Mount St. Peter near Maastricht (at 170 meters, this hill is hardly a "mount," but in a flat country like the Netherlands one has to lower one's elevational standards). Having collected several dozens of immature males and females from between pieces of agricultural equipment parked in the caves by local farmers, he then took them to the cold, damp, dark basement of the zoology lab in Leiden. There, under quasi-cavernous conditions, he hoped to induce the virgin spiders, after their final molt into adulthood, to mate while he watched them through his microscope under the illumination of a tiny lightbulb.

To his surprise, Van Helsdingen discovered that mating couples of this spider blatantly ignored the arachnological handbooks. Shivering over his microscope in that dank cellar, he saw that a male spider, once copulation was imminent, would *not* fill his pedipalps with sperm, but instead would embark on a lengthy period of repeatedly and swiftly locking his still empty pedipalps onto the female's epigyne, the complicatedly shaped entrance to her uterus. In the beginning, Van Helsdingen noticed, there would be some clumsy flubs where a male's pedipalp would slip off the female's epigyne, but after some time the male usually would get the hang of it. Each time, a pedipalp would remain locked on the female's epigyne for a few seconds only, followed by a few more seconds of cleaning and resting, and the male would tirelessly continue with these "dry" intromissions for hundreds or even thousands of times, a feat lasting up to six hours! By interrupting the coitus, isolating the female, and waiting for her to lay eggs—all of which proved unfertilized—Van Helsdingen confirmed that indeed no sperm was transferred during this phase.

After the drawn-out dry phase, the male would abruptly walk off and, at the edge of the female's web, start constructing a minuscule triangular sperm web on which he would deposit a drop of sperm, suck it up into both of his pedipalps, and eagerly return to his female to start copulating "for real," pushing his pedipalps into her epigyne and emptying them.

For years, Van Helsdingen would ask his colleagues at spider conferences whether they had ever observed the same, but nobody had—or rather, they'd all assumed that when any male spider started mating, he had already filled his pedipalps with semen beforehand. Only gradually, as more biologists began observing the spider deed with a critical eye, did it become clear that a protracted "dry" phase is a common feature of copulation among all sheet web spiders and a few other spider families as well. The sierra dome spider (*Neriene litigiosa*), for example, is a North American sheet webber that copulates in a way very similar to Van Helsdingen's cave species. Evolutionary biologist Paul Watson of the University of New Mexico has been studying these spiders—named for the dome-shaped webs they build among branches—at his field station on the shore of Flathead Lake in Montana for more than three decades, taking investigations of dry and wet spider sex a step further.

Watson placed mating spiders in an instrument called a respirometer,

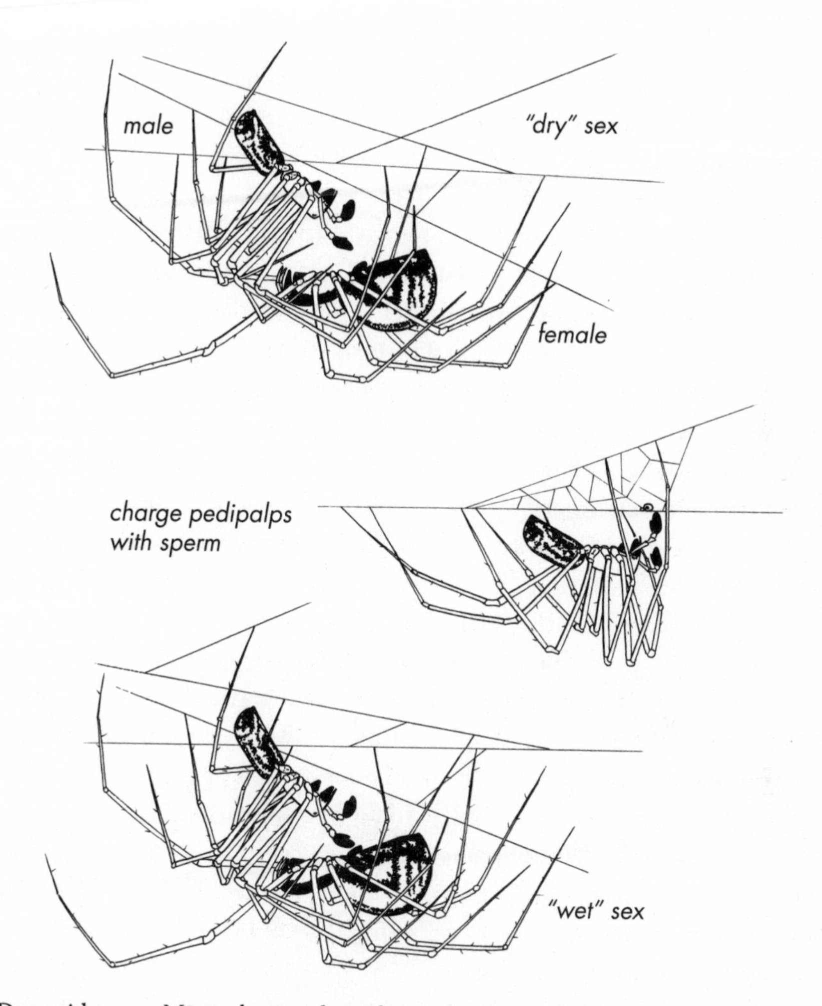

Dry spider sex. Many sheet web spider males first copulate with empty pedipalps, then run off to fumble with a droplet of sperm and a specially fashioned sperm web before returning to the female to inseminate her.

which accurately measures the amount of energy that an animal expends. As it turned out, during a bout of frantic dry copulation, a male *Neriene* spider would reach an energy output of about a tenth of a milliwatt. To the average lightbulb buyer, that may not sound like much, but, as Watson writes, "[T]his is about as hard as a spider can work"—literally as much work as running at full speed on all eights. This is due to the fact that locking the pedipalp onto the epigyne requires the laborious inflating of the so-called hematodocha. To do this, the male has to force blood into this large fluid-filled sac inside the pedipalp, which makes it unfurl like a party hooter. This then allows various mobile parts of the complex pedipalp to expand, slide, and penetrate into corresponding spaces provided by the female genitals. Watson found that the faster a male inflates, deflates, and reinflates his pedipalps (which ranged from once per minute to more than ten times per minute), the higher his energy output—and, possibly, the more desirable he would be as a mate.

Watson also employed a kind of genetic fingerprinting to work out which of the two to five males that a female normally mates with during one season would father most of her spiderlings. He found that although the first male would normally be the most successful, the success of any subsequent male depended to a large extent on his "copulatory vigor," a combination of the rate of dry intromissions, the number of flubs, and the duration of the dry mating phase. In other words, a female sierra dome spider chooses, from among the sperm she receives in a season, those that were donated by the males that stimulated her epigyne in the most energetic and memorable way. And one way in which she can exert such a choice is by deciding not to stick around for the wet phase and make herself scarce while the male is busy filling his pedipalps, thereby denying him his chances of fathering any of her eggs, despite his many hours of copulatory courtship. In many other animals where dry precedes wet, making a move in the lull between both phases is also a very effective way for a female to express her choice.

This New and Useless Part

When you think about it, a copulating pair of sheet web spiders is a powerful, though somewhat drawn-out, advertisement for the supremacy of cryptic female choice. If we did not know better, we might view the pair with Victorian eyes and see a male who has chosen his female and is now busy exercising his successful-suitor's rights. But in reality, it is the male who is dancing to the tunes of the female. He is wasting many hours of his short life on vacuous copulation with empty pedipalps, even though he would much prefer (if his spider brain were capable of such sentiments) to charge his pedipalps with semen *before* approaching the female and simply inseminate her immediately upon their first encounter. The fact that he, and many other male animals, have been forced by sexual selection to go through a lengthy copulatory process in which ejaculation seems not to be the be-all and end-all shows the extent to which his reproductive fate is in the hands of his female. Still, one could argue that even if a female might run away before he is through, or she may dump some of his precious sperm, these risks do not justify such extravagances as extended periods of useless, dry pedipalp thrusting. Or, to put it more generally, why don't male animals stick to a simple hit-and-run (or penetrate-and-ejaculate) strategy? That way, the time saved could compensate for those females that in the end do not use their sperm.

The reasons for this lie in the fact that there is female plumbing to consider. It is not just that premature ejaculation lowers a male's chances of fertilizing his female's eggs. In many organisms, it literally *ruins* his chances. This is because there are few kinds of animals in which males dump their sperm directly onto a female's eggs. In the vast majority of animals, the male's ejaculate only reaches a staging area to what may be a labyrinthine female system of valves, locks, narrow corridors, sluices, and blind alleys. In humans, for example, the male leaves his sperm cells in a puddle on the uterus's doorstep, after which they have to be transported up through the mucus-clogged cervix, along the walls of the womb, and through the valve-like entrance to the one fallopian tube that carries a fertilizable egg. They stand a chance at fertilizing that egg only if they make it all the way up to the ampulla, the sharp bend in the fallopian tube close to the ovary. And all along the way there are outward-directed microscopic hairs, narrow passages, and other obstacles.

Finding one's way through the reproductive maze inside a human female is still a breeze compared with the odyssey a sperm cell has to undertake through the piping of, say, the female hamster, in which the narrow fallopian tube opens next to a blind alley in the back of the uterus and is folded up into more than fifteen zigzag loops. In featherwing beetles and certain banana flies, the female tubing is similarly exaggerated. Or, in fact, in sheet web spiders, where the tunnels leading from the epigyne to the sperm storage site are sometimes ridiculously long and convoluted, consisting of two stacks of up to ten loops. This drives home the fact that

Labyrinthine plumbing. Females of many animals have their eggs hidden behind extensive systems of coiled tubes and valves. Here: (A) featherwing beetle; (B) banana fly; (C) sheet web spider; (D) exploded view of the stacks of coils in the latter.

....

insemination is not fertilization. If you're a male intent on reproducing, ejaculating into a female is only your first step. For your sperm to be actually guided through the maze that is your female's inside, you are going to need her help.

And a female may open her internal doors to a male's sperm only if the stimulation he has given her during the dry stage of copulation was acceptable to her—a fact that frequently frustrates artificial insemination in livestock. As those working in the "veterinary fertility" business know, a sow's response to a polyethylene pipette is decidedly lukewarm compared with a real male pig with real genitalia—and, crucially, she allows much less sperm to enter her uterus if it was administered by a man in a lab coat than by a proper boar. Artificial insemination efforts in pigs therefore resort to using real boars for mood enhancement and a latex pipette shaped to have an uncanny tactile resemblance to a boar's penis (sometimes with a special vibrator attached to it)—and indeed reap the reproductive benefits of this bit of swine sensory theater.

So an important incentive for a male to use his penis as an internal courtship device is to persuade the female to open inner doors that remain shut to lesser males. Biologists have discovered that the female reproductive system in most animals, great and small, carries one or more such valves, the operation of which is under unconscious female discretion. In mating spotted cucumber beetles, for example, the male rhythmically strokes the female with his antennae while he has his aedeagus inserted into her. All the while, though, the female keeps the muscles in her vagina taut, so that his aedeagus cannot penetrate all the way to her sperm storage site. Only when the female decides that the male has stroked fast enough will she relax her vaginal muscles for him. If not, the male will stroke endlessly and eventually dismount in disgust. In this section, however, I will focus on mammals, since the inner workings of their female genitalia are usually better known than those of other animals. Moreover, it will give me the chance to keep my promise of Chapter 3, to come back to the female orgasm, which may be crucial in the opening and closing of such reproductive valves.

You might be forgiven for thinking that hype and controversy surrounding the female orgasm and, by association, the clitoris are limited to the crowded layout of modern glossy magazines. However, the scholarly literature on the subject is a similar minefield of heightened sensitivity, hype, and hyperbole. In the mid-sixteenth century, the heated exchange between anatomist Gabriele Falloppio, who claimed to have "discovered" the clitoris (discovered for male-dominated science, that is; women had been aware of it since the dawn of time, of course), and the great Vesalius, who retorted that this "new and useless part" was likely found only in hermaphroditic freaks, is indicative of the roller coaster that would be the next 450 years of discourse on the clitoris.

For much of the seventeenth century, anatomists considered the clitoris as little more than a cue for them to draw their clitoridectomy tools, under the misapprehension that, as the authoritative Vesalius proclaimed, it was something that oughtn't be there on a woman. But around 1672, Dutch physician Reinier de Graaf published a detailed account of the full extent of the clitoris (and we shall see below how large this extent really is), exclaiming, "We are extremely surprised that some anatomists make no more mention of this part than if it did not exist at all in the universe of nature." He was also rather lucid in stating that obviously nature must have provided women with such a pleasure spot; otherwise why would they ever run the risk of pregnancy and labor?

With De Graaf's rediscovery, things were looking up for clitoral awareness—if it hadn't been for the fact that De Graaf soon took his own life in a fit of depression over an anatomical dispute with his rival Swammerdam. Its main advocate gone, the clitoris sank into medical oblivion again for more than 150 years. It resurfaced only in 1844, in the brilliant book *Die männlichen und weiblichen Wollust-Organe des Menschen und einiger Säugetiere* (The Male and Female Lust Organs in Humans and Some Mammals) by the German anatomist Georg Ludwig Kobelt.

Even in the late twentieth century, history repeated itself in a final (for now) cycle of discovery, oblivion, and rediscovery. When De Graaf and Kobelt described the human clitoris, their dissections revealed that the well-known, and usually minute, button-like glans and the elbowed ridge of the clitoris immediately underneath are really just the tip of the iceberg. Most of the clitoris is hidden away by pelvic fat and bone but is nothing to be sniffed at: two thin, 10-centimeter-long (4-inch) stalks branch out toward the buttocks and embrace two shorter, thick, pear-shaped "bulbs" that lie on either side of the deeper part of the vagina. The four together form a

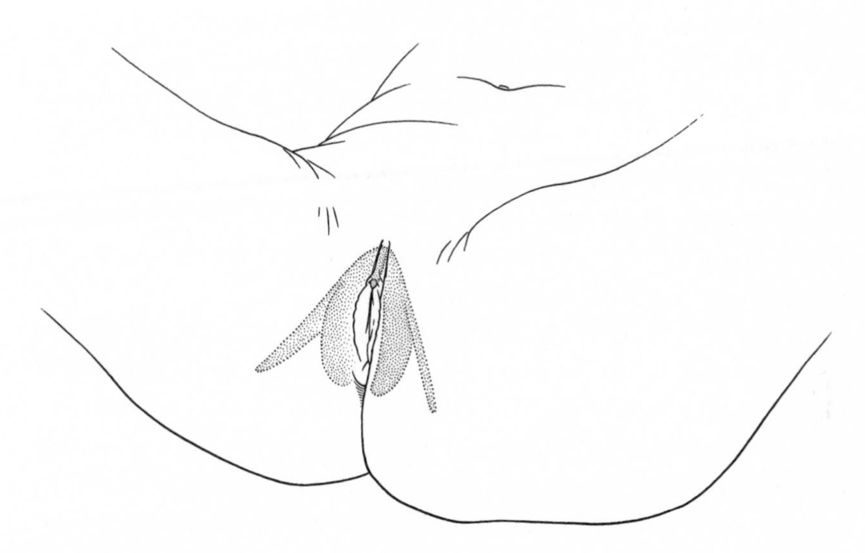

Not so puny. The true size and complexity of the human clitoris have been forgotten and rediscovered several times over the centuries.

large pyramid, only the acme of which, the clitoris proper, is visible on the outside. The whole organ is richly supplied with blood vessels to make it swell, and wired with incredibly thick bundles of nerves, of which Kobelt wrote: "Here they are, even before their entrance so very thick that one scarcely imagines how such an abundance of nerve mass can still find room between the countless blood vessels of this very tiny structure."

Despite these seventeenth- and nineteenth-century nuggets of accurate knowledge, until very recently many human anatomy textbooks either ignored the clitoris altogether or just briefly mentioned the minuscule externally visible bits. So, once again, it came as a surprise for many when, in 1998, Australian anatomist Helen O'Connell redescribed the full extent of the female clitoris using modern techniques. "Tens of times larger than the average person realises," *New Scientist* exclaimed.

Like its instigator, the female orgasm has met with a similar fate over the centuries. Men (it was always men) who were in the know either dismissed it as a vestige of our animal past or, as Freud did, viewed it as an infantile stage in female sexual maturation, to be replaced by proper adult vaginal pleasure later in life. Either that or they went completely the opposite way and glorified the female orgasm as a uniquely human capacity, evolved to enforce bonding between men and women. Such views, fashionable in the 1960s, were expounded by Desmond Morris, who wrote in his famous 1967 book *The Naked Ape* that the "female orgasm in our species is unique amongst primates," having evolved because of "the immense behavioural reward it brings to the act of sexual co-operation with the mated partner. Like all the other improvements in sexuality this will serve to strengthen the pair-bond and maintain the family unit."

The reality is different. Neither clitoris nor orgasm is uniquely human, both probably occur in most mammals, and the family unit does not enter the equation.

The existence of a clitoris in nonhuman mammals is the least controversial. In all mammals, the young embryo—in humans, until about the ninth week of pregnancy—has between its paddle-like hind limbs a small lump, the so-called genital tubercle. As the embryo matures, this tubercle then either extends at the top and grows into a penis in male babies or develops more at the bottom to form a clitoris in the females. Placental mammals like us, which develop in a comfortable, nourishing womb, are born with fully formed genitals. But marsupials, which are born much earlier in development and then hang around in the mother's pouch for a long time, come into the world at such an early state that the male and female genitals are still identical. To, say, a wallaby family, the answer to the question "Is it a boy or a girl?" only gradually becomes apparent with regular pouch checking and cannot be certain until about four months of pouch life.

But in marsupials and placental mammals alike, all females see their genital tubercles eventually mature into a full-blown clitoris, which can come in as many shapes and sizes as male mammals' penises. Ewes have a discreet and well-hidden one, while the clitorises of bonobos, certain marmosets, and other South American monkeys, rodents, and many carnivores are large affairs, with the external part bulging out like a small penis, often rigidified with an internal clitoris bone and sometimes supplied with bristles on the outside. As clitorises go, those of the mole and the spotted hyena really take the cake, being almost exactly the size and shape of a male penis, complete with a urethra passing through it and erection capabilities. The female spotted hyena in particular has taken life with a

17-centimeter-long (7-inch) clitoris to extremes, giving birth through it—the birth canal passes through it and the female often perishes in the process—and rolling it up like a sleeve whenever she copulates with a male.

Since in humans the clitoris is the seat of the female orgasm—so-called vaginal orgasms are thought to work via internal stimulation of the clitoris as well—it is likely that all female mammals experience orgasms. (In the next section, I will get around to explaining why this is important in the context of cryptic female choice—bear with me while I titillate you a little longer.) I say "likely" because it is hard to know for sure, of course. As behavioral biologist Tim Birkhead has lamented, "[H]ow do you tell if anything else is having an orgasm?" In their famous pioneering 1966 sexology book *Human Sexual Response,* William Masters and Virginia Johnson spent an entire chapter describing and dissecting the carnival of signs that betray "*la petite morte*" in a woman. They list an "increment of myotonic tension" in the long muscles of the entire body, "involuntary carpopedal spasm" (that is, clawing of hands and feet), rhythmic contractions of the anus, vagina, and uterus, as well as hypertension and hyperventilation. Since then, with improved technology, other features have been added to this list, such as a rush of the "cuddle hormone" oxytocin in the blood and a sudden change in activity of one particular part of the brain, the orbitofrontal cortex, as has shown up on MRI scans.

Of course, many of these "symptoms" can be triggered by other events too—a hyperventilating woman is not necessarily having an orgasm. Behold the quandary of the zoologist wishing to study orgasm in animals: while a human female can simply tell you and confirm whether or not she's had an orgasm, a female lab rat cannot. Still, zoologists have come up with some quite convincing observations that suggest that orgasm is rife in the female animal world as well. Those same lab rats, while not amenable to filling out questionnaires, have been anesthetized and rigged with recording equipment, which, after due stimulation of their genitalia, showed something called the urethrogenital reflex: a series of rhythmic contractions of their vagina and anus, the tracings of which cannot be distinguished from those obtained from humans in the throes of orgasm. And back in 1952, intrepid researchers Noland VanDemark and Ray Hays inserted little water-filled balloons, made from the thumbs of rubber gloves, into the uteri of cows in heat and allowed a bull to copulate with them while they recorded (from a safe distance, presumably) the pressure changes inside with an ink-writing lever attached to the balloon via a thin rubber tube. Although the bull's mount, insertion, ejaculation, withdrawal, and dismount took only five seconds, the researchers' pressure gauge showed "tetanic contractions of the uterus" for up to two minutes after the bull had dismounted. An orgasm? A moooot point.

In female primates, which may climax in a way a bit more easily recognizable to human observers, the results are less equivocal. In 1970, anthropologist Frances Burton of the University of Toronto took it upon herself to test the then widespread idea that female orgasm was something as uniquely human as language and tool use. She put mature female rhesus macaques, belly down, on an apparatus that can only be described as a bondage table, where they were strapped in dog harnesses and wired up to monitor heart rate. It was a rather cruel setup, but the only way to keep the monkeys in a position that allowed Burton to bring them to orgasm. Surprisingly, given the less than romantic conditions, the monkeys sometimes complied. After relaxing her test animals by grooming and feeding them, Burton proceeded to subject their vaginas to regular thrusting with an artificial monkey penis. In several cases, the female would at one point start to grunt, look back at Burton, and make clutching backward movements with her arm—a behavior that primatologists had already suspected to indicate the verge of orgasm in macaques. In a few cases, Burton observed series of "intensive vaginal spasms" very similar to those known from orgasming humans.

The Scotland Yard

The reason I am going on about the female orgasm is that scientists have repeatedly suggested that it is one way in which a female mammal can influence which males' sperm are allowed to travel up her reproductive labyrinth and which are not—a kingpin, if you will, in sexual selection in mammals. Remember that Robin Baker and Mark Bellis discovered that there was much less flowback of sperm if a woman had had an orgasm during or after her partner's ejaculation, and that by means of her orgasm a female may be able to "select" a particular male's sperm by giving it a push, as it were. What we did not get into is exactly how this would work. The answer, at least according to adherents of the so-called upsuck hypothesis, has to do with hydraulics.

In 1952 (clearly a golden year for low-tech veterinary reproductive experimentation), the scientist Ramsay Millar described in the *Australian Veterinary Journal* how he had connected (as one does) the uterus of a thoroughbred mare to a bottle of methylene blue via a copper tube and during mating saw how the uterus suddenly developed a low pressure, leading to some 80 milliliters (about 3 fluid ounces) of the blue liquid being sucked up. During the same period, other zoologists performed similar experiments on rats and mice and discovered that colored liquids, upon "manual stimulation of the vulva," were sucked into the uterus. In the late 1960s, the obvious next step was taken: to test this in humans.

Well, rather, in *a* human. In 1970, Dr. C. A. Fox and colleagues of the National Institute for Medical Research in London reported how they had planted a tiny electronic pressure sensor/transmitter in the uterus of a woman—presumably Dr. Fox's wife, Beatrice, who in the acknowledgments section of the article is thanked rather emphatically for her "help." She then engaged in heterosexual coitus—with the article's first author, one imagines—while the pressure changes in her uterus were recorded by a receiver placed under the mattress where the scientifically justified deed was done (twice). In both cases, the readout of the pressure gauge showed a sharp drop in pressure inside the woman's uterus immediately after she climaxed.

If the upsuck hypothesis is correct, then by having an orgasm female mammals could give the sperm of a more stimulating male an advantage in the sweepstakes. The fact that most female humans (and monkeys) do not have an orgasm each time they copulate seems to fit this idea, as do the discoveries that certain males induce orgasm in females more easily than others. For example, in 1990 Italian primatologists Alfonso Troisi and Monica Carosi studied sexual behavior in a large group of Japanese macaques at the zoo in Rome and saw that during copulation, females would sometimes clutch and look back at the male, show muscle spasms, and "vocalize"—the same kinds of behavior that Frances Burton had seen in rhesus macaques and that seemed to be a telltale sign of the macaque climax (assuming monkey females don't fake it). Troisi and Carosi saw this happening about 60 percent of the time when a low-ranking female was mounted by a high-ranking male, whereas in matings within the same social rank or when the female was down-dating, females would climax only once in five copulations.

The problem with such animal studies is, again, that it is hard to be sure about the females' orgasms, short of rigging each one with a set of sensors and transmitters in their genitalia. At least in terms of certainty about what is an orgasm and what is not, it may be much more reliable to work with organisms that will simply reply when questioned about their orgasmic experiences. Such as university students. In exchange for fourteen dollars or course credits, David Puts, Lisa Welling, and their colleagues at Pennsylvania State University got seventy heterosexual student couples to have their photos taken and fill out questionnaires about the last time they had sex with each other. The women were asked (out of earshot of their partners) whether they had had an orgasm and, if so, at what point during their coitus. Then the photos of the male students were sent off to a UK university, where a jury of nine men and nine women rated their good looks on a scale of zero to seven. The results showed that women who had had sex with a male partner who (at least in the UK) was considered particularly attractive had much more often experienced an orgasm during or shortly after the male ejaculated—roughly the interval that Baker and Bellis had found was needed for more sperm uptake.

And in another study, a Portuguese-American-Scottish research team had more than 320 mostly Scottish women fill out a questionnaire on orgasm and sexual preferences. They found that the majority reported that men with a longer-than-average penis—longer than a twenty-pound banknote, the researchers had helpfully added in their questionnaire—more frequently elicited a vaginal orgasm with them.

Still, not everybody is convinced that in humans orgasm is really a way in which females exert their cryptic choice from among the males that inseminate them. In fact, there is a school of thought that doubts that the female orgasm serves any function at all. Donald Symons, author of the 1979 book *The Evolution of Human Sexuality,* was the first to suggest that perhaps it is just a pleasurable but functionless vestige. Not in the Freudian sense, of a dark animal past, but of the developmental program that male and female babies go through during their first few weeks in the womb.

After all, anatomically, the clitoris and the penis are so-called homologues, organs with the same basic blueprint that grow from that same genital tubercle between the embryo's leg stumps. Not only that, but the nerves and hormones involved, yea, the whole urethrogenital reflex, which sets in motion those 0.8-second genital spasms during female as well as male orgasm, are the same in men and women. Perhaps, says Symons, women have orgasms simply because *men* have them. And men have evolved orgasms because they link pleasure to ejaculation and to copulation. For men, and males in general, it pays to be on the prowl for more matings. After all, since Bateman we know that more copulations mean more offspring for males, but not for females. So an orgasm reward mechanism that makes males pursue an ever greater number of sexual encounters would instantly be spread by evolution. And the female orgasm may just be along for the ride. Just as male nipples are a pointless by-product of the evolution of the infinitely more useful female mammae and teats, the female orgasm may be a by-product of the male's. As famed evolution writer and by-product enthusiast Stephen Jay Gould wrote in his essay "Male Nipples and Clitoral Ripples," male nipples and female orgasms exist because "males and females are not separate entities, shaped independently by natural selection [but] are variants upon a single ground plan." In other words, it is possible that men and women share a characteristic that has given evolutionary benefits to only one of them.

Given the evidence from uterine hydraulics in livestock, climaxing monkeys, and also tantalizing discoveries such as that the release of oxytocin during orgasm causes sperm to be transported along the uterus wall to the one ovary with a ripe follicle, I would not put much money on the by-product hypothesis. Still, the last word on this has not been said. Far from it: the by-product theory was rehashed with gusto in 2005 by Indiana University philosopher Elisabeth Lloyd—a former student of Gould's—in her book *The Case of the Female Orgasm,* which has single-handedly rekindled the whole discussion on the role of the female orgasm in mammals. It has also spurred some researchers into action to obtain new data. And, frankly, this is sorely needed.

For despite all the pages devoted to the biology of the female orgasm—in the past five years alone, there have been a whopping five hundred scholarly texts on it—the hard data that we possess are very few, especially where our own species is concerned. There was only a single Mrs. Fox who ever had a pressure sensor placed in her uterus, and those two 1970s pressure tracings are all the upsuck theorists have to show for themselves. Baker and Bellis got their information on semen flowback after orgasm from just eleven women, only one of whom was responsible for two-thirds of the semen-filled condoms that they analyzed. And questionnaires, even if the whole female population of Scotland were to fill them out, can only get you so far. To really understand what role the female orgasm plays in cryptic female choice, scientists have to start studying, either in humans or in lab animals (and with a sample size larger than one!) what impact orgasm has on the chance of bringing a sperm and an egg together. As Olivia Judson, author of the inimitable *Dr. Tatiana's Sex Advice to All Creation,* has written, "[I]t's time to collect data. Without it, the debate will remain like sex sometimes is: furious, empty and anticlimactic."

Female Larders and Fetal Loss

Every so often a story hits the headlines of a woman who impregnates herself with her ex-partner's sperm secretly kept in the freezer, and of all the unsavory legal and moral consequences that ensue. In the animal press—the *Daily Dung Fly,* the *Snail Mail,* the *Turtle Times*—such stories would hardly raise an eyebrow. In fact, their readers would probably shake their heads and drily remark how typical it is for *humans* to make such an issue out of this. For in most animals, females routinely keep souvenirs of past mates in the form of sperm samples stored in their reproductive system. And they do so for exactly the same reason as those human women and their freezers: one day they might like to have babies from long-gone ex-boyfriends.

We all think we know that sperm cells do not live long once they have been ejaculated and have landed in a woman's vagina. Their days in this new, feminine environment are numbered; in just three days, at the most four, they will expire, and if there is going to be any fertilizing, it has to occur before that time. But the fragile constitution of human sperm is actually an exception. In most animals, sperm inside a female can remain fresh for much longer. Bats in colder climes, for example, engage in a mating frenzy in autumn and then, sperm securely stuck against the walls of their uteruses, go happily into hibernation and only upon awakening in spring do they release the still sprightly sperm cells to produce baby bats. To some snakes and turtles, it's no skin off their reptilian noses to store a male's semen for several years, since suitable males are often few and far between. And in many an ant nest, the same queen rules the colony for decades, all the while fertilizing her millions of eggs with the only sperm she ever received: during her nuptial flight, right at the start of her reproductive career. Her sperm stores remain viable for all those tens of years.

So what do these female sperm larders look like, anatomically, and how do they function? There is great variation among different kinds of animals. Many snakes have deep lengthwise folds in their oviduct (the corridor leading from cloaca to ovary) where sperm are stockpiled and kept in suspended animation. Insects, on the other hand, have at the end of the vagina a central sperm chamber (often very large and convoluted) from which long tubes lead to one or more sperm pouches, where sperm can be kept until needed. In land snails, the layout is even more complex; as with many other mollusks (see Chapters 1 and 8), the sperm is delivered in a sealed package, or spermatophore, which is first taken up into the so-called sperm-receiving organ. Here, the sperm package is dissolved, and only a few sperm escape digestion and travel up the oviduct. There, they are led into one of a number of separate blind alleys, or "sperm tubules," where they can be used at will for fertilizing eggs at any time.

Although females throughout the animal kingdom can avail themselves of such a great choice of possible types of sperm repositories, these all share one important feature: they are ultimately under female operation. Males may deliver their sperm directly into a central sperm deposition chamber, but it is the female's muscles and nervous system that then transport them to specialized long-term storage organs, and all withdrawals of sperm savings from these larders are similarly under female control. And this sperm mobilization comes on top of the active sperm uptake that happens immediately after ejaculation, which, as we saw in the previous section, in mammals may be brought about via orgasm.

What is important to note is that there is often not just a single long-term sperm store, but many separate stashes. Many flies have two or three separate pouches in their "spermathecary"; the edible escargot *Cornu aspersum* has between three and nineteen separate sperm stores, and turtles have countless minuscule sperm-harboring tubes all along their oviduct.

You may have already grasped the implication of this—it means that, in principle, it may be possible that a female stores sperm from each male she has mated with in its own, distinct "freezer drawer" and can summon a particular male's sperm at will.

I say "in principle" because the evidence that this actually happens is still rather weak. The best indications of such selective sperm use by females come from a rather unlikely source: fresh cowpats in meadows. Or rather, the dung flies that call such places home.

The yellow dung fly (*Scatophaga stercoraria*) is a very common sight in the Northern Hemisphere, especially in pastures and meadows where cattle and sheep provide a constant supply of fresh dung. The flies themselves hardly eat the dung (they mostly feed on nectar and smaller insects), but their larvae do, which is why, to a dung fly, a fresh cowpat is one of the most romantic places imaginable. It is where they court, mate, and lay their eggs. Cruising low above the grass, the flies drop themselves to the ground as soon as their antennae register the unmistakable scent of newly produced manure, and, since they often overshoot their target, find their way to the dung on foot. Once there (or even on the way there), males seize any females they encounter and try to mate with them. They do not care too much if another male is already doing the same, so by the time a female has reached the surface of a cowpat, she is often tended by two or more males. The later-arriving males try to squeeze in under the male that has already mounted her while he, holding on to his prize with his front legs and balancing himself on the dung surface with his hind legs, uses both his middle legs to try to kick his competitors away. Often the female is the one who suffers most during such a tussle, and many a female literally drowns in cow poo under the pressure of her many suitors. So perhaps "romantic" isn't really the word after all.

It was an English ecologist, Geoff Parker, a former classmate of Robin "Flowback" Baker's, who first turned these much ignored golden flies into heroes of sexual research. Completing his doctoral studies at Bristol University in the late 1960s, he realized that dung flies on cowpats were in many ways the ideal system to base one's behavioral ecology Ph.D. on. First of all, the work was inexpensive: he needed only "a stopwatch, ruler, thermometer, tape recorder, glass vials, entomological pins, a notebook, and a pencil." Second, the flies and their habitat were ubiquitous; doing his fieldwork meant little more than stepping into his local cattle field, gently draping an old coat side-wind to the excrement, and gathering data. The only real peril was "the occasional undetected bull," and the only irritations "the rain and the curiosity of unhabituated cows, dogs, and small children." So while his fellow Ph.D. students went off on costly expeditions to observe large game in Africa and came back after many months with only a few hours of observations on a single cheetah, Parker spent his summers propped on an elbow next to a cowpat quietly amassing a volume of work that has since become a classic in ecology. And in terms of the drama of the sexual behavior he observed, the flies were not outdone by big mammals. As he wryly observed, "[I]f dungflies were the size of red deer they would be the subject of a thousand books and nature films."

But as dung flies are only the size of dung flies, and what with their scatological tendencies, they adorn only the pages of niche journals like *Ethology, Evolution,* the *Journal of Evolutionary Biology,* and *Behavioral Ecology and Sociobiology.* And have done so in an uninterrupted series of scientific publications produced by Parker and his students for the past forty-five years. One of the things they found out is that a female dung fly possesses one large central sperm chamber, the "bursa copulatrix," and three smaller "spermathecae," each one connected to the bursa via a long, narrow duct. Two of these spermathecae lie on the right-hand side of her abdomen, while the third one lies on the left. In 1990, Parker's former Ph.D. student Paul Ward set up a dung fly lab at the University of Zurich in Switzerland (where the hills are alive with the hum of dung flies). And in the twenty years up until his untimely death in 2010, Ward's work showed that female dung flies can play these internal bagpipes with amazing finesse.

In the lab, Ward let female dung flies copulate for twenty minutes with one or two males. Then some of the females were allowed to lay their eggs on nicely measured bits of cowpat in petri dishes, after which Ward checked which larvae were fathered by which males; this was possible because the males had been bred to contain different gene mutations, by which their offspring could be recognized. The other females were killed after mating and then dissected so that Ward could count the number of sperm cells in each of the spermathecae. What he found was that a female prefers to store more sperm from a large male—which Parker had shown to be more attractive to females—than from a smaller male. She also stores large-male sperm preferably in one or both of the spermathecae on the right, rather than spread evenly across all three, as she tends to do with small-male sperm. And Ward also discovered that when it's time to lay eggs, the female would preferentially squeeze sperm from her right-hand spermathecae, so that the big guy would father most of her babies.

Later, when genetic fingerprinting techniques were developed for dung flies, Ward's team was even able to extract the sperm from the separate spermathecae and show that, indeed, these were from different males in different proportions. That all this sperm positioning was mostly the female's work, and not influenced by the male, was proven in a cute little experiment done by two collaborators of Ward's, Barbara Hellriegel and Giorgina Bernasconi. They divided just-mated females into two groups: some were left awake while others were put under anesthesia with carbon dioxide. They found that only the conscious females were able to create

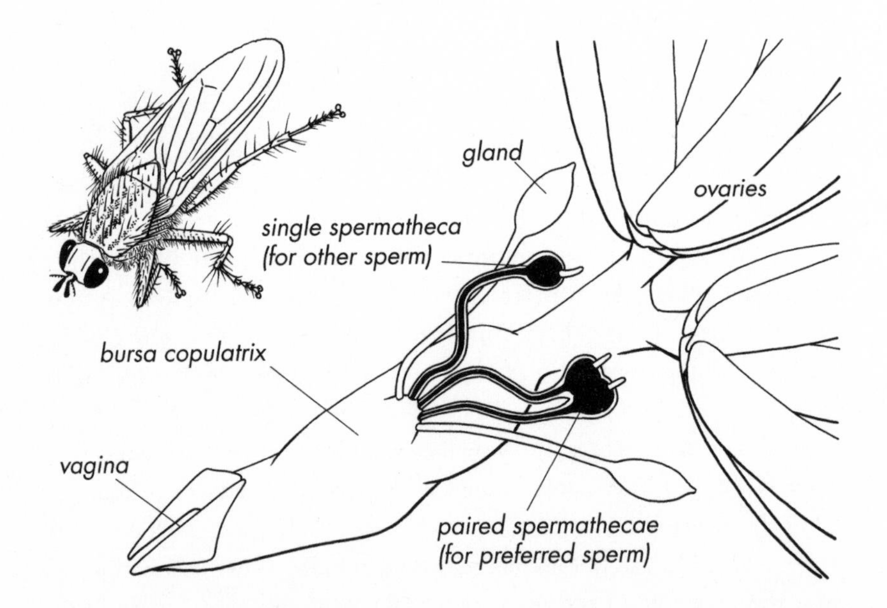

Female discretion. Internally, a dung fly has three spermathecae for keeping sperm from the males she has mated with. The paired spermathecae are reserved for "sexy" males; the single spermatheca is a backup for sperm of lesser quality.

....

separate sperm stores, while the ones under narcosis spread the sperm randomly.

So the yellow dung fly is the last in a long series of organisms that teach us that females in the animal world have a wide variety of tricks up their sleeves to determine which sperm is going to get access to their eggs. Not only can a female enforce dry copulation on her mate and give him the slip before his apogee, she can also dump or eject his sperm, or use internal valves and locks to keep unwanted sperm away from her eggs. Even when she accepts a male's sperm, she can store it away in a safe location and decide later whether or not to use it. And the list does not end there. Even after they have fertilized their eggs with a male's sperm, some females can, as a last resort, deny their embryos the opportunity to come to term. Such drastic measures, used by some female mammals to favor certain males over others, are known as the "Bruce effect." Rather disappointingly, the term does not refer to a disinclination in women to have their babies fathered by the kind of men who are called Bruce. Instead, it is named after Hilda Bruce, the English zoologist who revealed it in 1959.

The dramatic phenomenon was discovered quite by accident. Bruce, who was studying the effect of hormone supplements in lab mice, noticed that a pregnant female mouse would sometimes miscarry when she was introduced to a new male, and would instead conceive an entire new litter from this novel mate. Since then, zoologists have found the Bruce effect in many other lab animals, especially rodents, and the cue from the new male that induces the female to give up on her existing pregnancy differs quite a bit. In the woodland vole, *Microtus pinetorum,* for example, the trigger is in the male odor, but in the field vole, *Microtus agrestis,* mere physical contact with a strange male can induce abortion. In the meadow vole, *Microtus pennsylvanicus,* finally, a female would abort and reovulate only once she had actually copulated with a new male.

For a long time, the Bruce effect was seen only under very unnatural laboratory settings, and many a zoologist wondered aloud whether it was a natural phenomenon at all or just a quirk brought about by the highly artificial conditions in a laboratory. But then, in 2011, it was also discovered in the wild. Zoologist Elia Roberts, who spent many years studying wild troops of geladas (a relative of the baboon) in Ethiopia, found the telltale sign of a natural Bruce effect in her geladas' feces. By measuring the level of estrogen in the droppings from particular females, she could tell if they were pregnant, and, when a sudden and permanent drop in the estrogen level occurred, that they had aborted. In almost all cases where there was such a premature abortion, this happened immediately after the troop of the female had been taken over by a new alpha male.

Now, the Bruce effect as we know it is probably an adaptation to the sad reality of infanticide. The males in many of the species in which the phenomenon has been found have the habit of killing any offspring of their mates that wasn't sired by themselves. In geladas, whenever a new male ousts the resident alpha male, the first task he sets himself is to execute all the juveniles fathered by his predecessor. So if the new male is going to kill her babies anyway, a female can invoke the Bruce effect to save time and energy wasted on a doomed pregnancy.

However, the Bruce effect might also signify a more general ability in female animals to terminate pregnancies at will. In many mammals, miscarriages are very common. In humans, it is thought that up to two-thirds of pregnancies abort spontaneously, often without the woman being aware that she was even pregnant. The majority of such abortions are probably due to the female body screening for serious genetic defects in the fetus, but there are more subtle mechanisms. For example, both pig-tailed macaque and human females miscarry more often if their mates have immune systems that are very similar to their own. Again, this is probably a strategy evolved to maximize a baby's disease-fighting ability—inheriting different immunity tools from both parents is better than inheriting the same set from both—and it is also not clear whether such abortions are triggered by the male or by the fetus. Still, these are intriguing indications that females could summon abortion as a last-resort effort to deny certain males their reproductive chances. (Counterintuitively, the phenomenon is misappropriated by some antiabortionists with the tortuous reasoning that if a woman's body can abort naturally, we should not do it artificially.)

To sum up, the female is anything but a passive sperm receptacle that nurtures the life breathed into her eggs by the males that dignify her with their semen. Instead, her body harbors a varied arsenal of ways to peeve her partner. It is a highly sophisticated engine of preference that springs into action upon each coitus. Taking cues from the male and his "courtship device," it winnows semen, rejects or selects sperm, and works its internal valves and springs like a mail-sorting machine. As we shall see in the next chapter, it is this intimate interaction between male and female genitalia that is the driving force for their unpredictable evolutionary trajectories. But we shall also see that there are limits to what this force can achieve.

Chapter 5

A Fickle Sculptor

Roughly speaking, biologists come in three flavors: field, lab, and theoretical. Those biologists who are happy only when they can don a raincoat, put a pencil and notebook, binoculars, hand lens, and collecting jars in an old military bag, and walk off into the wild clearly belong to the former kind. Lab biologists find the great outdoors a much too confusing place and prefer to parcel biological processes into nicely contained simple systems in petri dishes. And then there are the theoretical biologists, for whom even a fruit fly in a lab bottle is way too unpredictable. They forsake real life altogether and instead capture it in formulae on paper and strings of computer code.

Famed sexual selection researcher Andrew Pomiankowski of University College London snugly fits in the latter category, having authored papers with titles like "The Costs of Choice in Sexual Selection" (1987), "Why Have Birds Got Multiple Sexual Ornaments" (1993), and "A Resolution of the Lek Paradox" (1995)—all highly influential but also highly devoid of actual animals or observations. Instead they are chock-full of formulae and computer simulations, with opening sentences like "Let t be a male trait used by females in mate choice and p be the strength of female preference." Still, even theoretical biologists understand that the market value of their theories increases if they can be applied in the lab and in the field, which is why, since the late 1990s, Pomiankowski has been testing his work willy-nilly on animals in the wild. And that explains why, several years ago as I lived and worked in Borneo, I received an e-mail from him asking me to take him a-hunting for stalk-eyed flies.

Stalk-eyed flies, or Diopsidae, are among the weirdest insects you can find in the Tropics. Immediately after emerging from the pupa, when the outer casings of their bodies are still soft and pliable, both males and females (but especially males) inflate and extend the struts that their eyes sit on, and these harden into enormously elongated stalks, which they then fly around with for the rest of their lives. It is the culmination of aeons of females preferring males with eyes that stand far apart. In the most extreme cases, such as the species *Teleopsis belzebuth* from Borneo, the flies have evolved an eyespan that is a whopping two and a half times the length of the body—headgear that is the insect equivalent of the peacock's tail: cumbersome, extravagant, and very much sexually selected.

As in peacocks and other pheasant-like birds, mating in stalk-eyed flies takes place in so-called leks (the Swedish word for "game" or "play," Scandinavia having been the epicenter for sexual selection research for many years). Leks are a kind of love-in where groups of males and females congregate and where the latter choose from among the former and copulate with them. Stalk-eyed fly leks crop up at dusk on the tiny twigs and root hairs that stick out from the eroded banks of streams in tropical forests. So as the sun was setting over Malaysian Borneo one late afternoon in April 2006, I found myself clutching the spongy stems of wild yams and stumbling over loose branches and boulders, while slip-sliding down a steep slope in the company of Pomiankowski and my friend the tropical ecologist Stephen Sutton. We eventually reached the stream and, still a little unsteady, began aiming our headlamps at the muddy rootlets that Andrew had said might contain the fly leks. And so they did: here and there, congregations of the nimble flies revealed themselves, jerkily moving up and down their root hairs with their ridiculously wide eyestalks sticking out on either side, like tightrope artists carrying balancing poles.

The work of Pomiankowski and his collaborators has shown that each lek is controlled by a single male. Females prefer to alight on root hairs tended by the males with the largest eyespan. Laboratory studies proved that the eyespan a male can attain is genetically determined but that it also says something about the male's quality as a mate: males can develop large eyespans only if they have coped well with the toils of larval life, and long-eyed males also have larger testes that produce more sperm, and so are able to fertilize more eggs in one copulation.

As we saw in Chapter 3, when we talked about the sexual selection by females of males with exaggerated adornments, this amply explains why eyespan should keep increasing in a stalk-eyed fly species. Males with longer eyestalks get to mate more often and thus sire more offspring, and those offspring inherit not only the genes for long eyestalks from their fathers, but also, from their mothers, the genes that make females like long eyestalks—leading to the species snowballing evolutionarily into ever greater stalk length. But there are complications, and these are the ones that Pomiankowski has been tackling in his computer models. For example, why do some species have longer eyestalks than others? And what happens when eyestalk evolution finds itself in the dead-end alley where all males have the longest possible eyestalks and all the females have maximum preference for this, an evolutionary gridlock known as the "lek paradox"?

What Pomiankowski, together with his colleague Yoh Iwasa from Japan's Kyushu University, was able to work out is that such sexual selection is not a one-way street. In the evolutionary history of any stalk-eyed fly species, there may come a point where all males have the maximum attainable eyespan—any longer and they will no longer be able to fly or their eyes will simply snap off—and all females find this as thrilling as their tiny fly brains are capable of. But once this state of affairs is reached, the benefit of choosing disappears, since all males have become the same. This means that evaluating and choosing males– the whole business of lekking—has now become a waste of time and energy for the females. The result is that mutations that make females *less* choosy will suddenly confer an advantage, and, as Pomiankowski and Iwasa witnessed on their computer screens, such a population could actually start evolving to have shorter and shorter eyespan. In fact, they found that once they simulated several genes for male eyespan and for female preferences, a population would never stabilize and would continue to bounce around the edges of eyespan and eyespan preference space for the rest of its evolutionary life.

In this evolutionary restlessness lie the greatest differences between natural and sexual selection. Natural selection, where a species adapts to, say, soil type or temperature, has a single optimum. Soil and temperature tend to stay more or less the same over long periods of time and do not change in response to organisms that adapt to them. There are no feedback loops, so over many generations a species will inch closer and closer to the best fit. But sexual selection is completely different. There is no single optimum that the species is evolving toward. Instead, the male half of the species is adapting to the female half and vice versa. The fact that both genders are tracking a moving target is already enough to guarantee perpetual evolutionary motion, and the combining, mixing, and redistribution of male- and female-adapted genes in each generation adds another layer of complexity. So sexual selection is the acme of evolutionary dynamism—and therefore much more complex and hard to predict than natural selection.

Pomiankowski's number crunching to figure out the evolutionary tango of male and female stalk-eyed flies is just one example of a process in sexual evolution so complex and hard to predict that we need computer simulations to understand it. Another example is a phenomenon called the "rare-male effect." In fish and insects, and probably in a variety of other animals, too, females sometimes prefer to mate with the most "unusual" males.

Take guppies. As anybody who has ever kept these popular aquarium fish knows, guppy males come in an amazing range of genetically determined color patterns, made up of blotches and stripes of yellow, red, and black as well as patches of golden, green, or purple metallic sheen. And although in an aquarium you may find many different color forms in the same tank, in the Central American streams that are the native home of these fish, male color patterns are to some degree partitioned by watershed. In Trinidad, for example, the La Selva River houses a type of male prosaically called M7, which sports bright gold vertical bars on the base of the tail, a black spot at the base of the tailfin, and orange rims on the top and bottom of the tailfin. The M1 males from the nearby Guanapo River, on the other hand, have a white dorsal fin, a black bar on the tail base, and orange spots on the bases of the fins.

In an experiment carried out at the University of California, Riverside, researchers showed groups of M1 or M7 males to virgin female guppies from the Guanapo River. They set up guppy peep shows by dividing their fish tanks into male and female compartments with glass walls, so that the females could see (and be courted by) males, but could not actually mate with them. Then they took individual females out of the tanks and placed each of them with one M7 male and one M1 male and left these trios to sort out their sexual preferences in privacy for twenty-four hours, after which the female was removed and placed in her own aquarium to give birth to her young. Since the color patterns are genetically based, the researchers could easily see how many of each female's fry had been fathered by the M7 male and how many by the other. The result was that females that had previously been around M7s were about three times less likely to have their eggs fertilized by these males than females who had never seen an M7 before. Familiarity had bred contempt.

This rare-male effect, where females seem to prefer to mate with the new kid in town, is not unique to guppies. It has also been reported in many kinds of insects, where females prefer to mate with males with, for example, uncommonly white eyes (banana flies) or peculiar styles of courtship (crickets). But it is not entirely clear why females should so love a stranger. It might simply be the kind of sensory drive that we saw in Chapter 3: a female's nervous system will fire up more eagerly when it is suddenly confronted with new signals it has not yet been numbed by. But perhaps more often it is a mate choice strategy that has evolved to prevent inbreeding: too much mating with relatives is dangerous for the offspring's genetic health, so a preference for males that look as if they are not from around the neighborhood may be an advantage.

Either way, the rare-male effect might also lead to the kind of evolutionary dynamism I mentioned above. And again, it has required a theoretical biologist to fully reveal this for us. Hanna Kokko, a Finnish theoretician at Australian National University (known for the mysterious haikus with which she summarizes her research projects on her home page: "Bird here, two out there / Eggs in baskets wet or dry / Covariances"), has been the first to run computer simulations of the rare-male effect. Her work revealed that in a population of, say, guppies, where several color types of males exist, as well as a rare-male effect in the females, the genes for different male colors will keep fluctuating in frequency over time.

This is because a rare type of male will, as time goes on, initially become more frequent as it enjoys female preference. However, in this success lies its downfall, for the commoner the male type becomes, the less it will be considered sexy by rare-male-preferring females. Eventually, female attention shifts to the next rare kind of male. At the same time, the genes that make females prefer rare males are also prone to such fluctuations: whenever these genes increase in the gene pool, rare male types will rise to commonness in just one generation, rendering their offspring unattractive for not being a rare type anymore. And since these offspring also carry their mothers' preference in them, these genes, too, will decline in the next generation.

What the computer simulations of guppy color as well as the sexiness of fly eye length show us is that when male signals and female preferences evolve in a species, there may be feedback loops that could cause unpredictable patterns. Like weather systems, where humidity, air pressure, and sunlight all conspire to drive chaotic cycles of rain and shine, the evolution of seduction is full of unpredictability. Imagine that you have a set of identical animal species and let sexual evolution run its course in each of them. It would be like setting up a number of PCs and letting each of them run a simulation program written by Kokko or Pomiankowski. Even if the initial conditions (the population sizes, the length of a generation, and the kinds of genes responsible for signals and preferences) are precisely identical at the outset (which in nature they will never be), the random effects of demographics and genetic mutation will cause these runs to branch out in different directions in just a short span of time. When the simulations are over, the sexual signals and preferences in each species—initially identical—will have drifted apart beyond recognition.

Now substitute genitalia for color patterns and eyestalks. If sexual selection on genitalia indeed progresses in a similar fashion, as Eberhard believes, then evolution will be like a fickle sculptor with ever wet clay, molding genital shape in continuously changing shapes and forms, never satisfied, never allowing it to set. And with such rapid and erratic evolution, it is no wonder that different species have such different genitalia, since each is just a snapshot of that ongoing dynamic interaction among genetic mutation, sensory drive, and female choice.

But it is easy to get carried away by computer simulations and fancy evolutionary theory. Is this really what goes on out there in the wild world of real animals with real genitals? What do we see when we try to watch genital evolution in action? Time for a reality check.

Look Back in Amber

Observing ongoing evolution of genitalia in the real world is not easy. Other organs and body parts have been seen to change shape in wild animals over periods as short as a few decades (the beaks of Darwin's finches in the Galápagos Islands, for example), but as far as I know, nobody has accomplished the same with genitalia. Fortunately, paleontology provides us with the next best thing: a peek into the procreative past.

As Australian geologist John Long explains in his 2012 book *The Dawn of the Deed,* the paleontology of sex is still in its infancy—a fact largely due to the puny likelihood of sexual organs, let alone sexual behaviors, fossilizing and making it to museum drawers. There are a few exceptions, and Long describes them lovingly. Like the three-hundred-million-year-old pair of sharks from a Montana sediment, one biting the head spine of the other, a preliminary to shark sex (sadly, in this case, never consummated). Or the famous fossil pit site of Messel, in Germany, which has yielded many petrified mating pairs of turtles, each couple cemented in eternal embrace. And Long himself has been instrumental in the discovery of the world's oldest pregnancy and umbilical cords in so-called placoderm fishes from the Devonian, 380 million years ago, and working out the mechanics of their copulation and reproduction.

Less ancient, but much better preserved, are those insects caught in the act by a clear drop of ancient pine tree resin, hardened into the famous twenty- to forty-million-year-old amber of, for example, the shores of the Baltic Sea or the Dominican Republic. Prized pieces in collections of insects in amber often contain entombed mating pairs of fungus gnats, scavenger flies, or mites. And the nice thing about these insect copulas is that they are not just hollow shells; the genitalia that once were central to their erstwhile sex lives still reside, often perfectly preserved, inside the amber sarcophagus. The only problem is that in order to study those internal organs, the valuable specimen needs to be destroyed. You'd think. But Michel Perreau, a physicist/entomologist at University Denis Diderot in Paris, has found a way to take a peek at those genitalia without needing to crack the specimen.

Using a technique developed for medical imaging at the particle accelerator of the European Synchrotron Radiation Facility in Grenoble, Perreau managed to make high-resolution CT scans of a tiny beetle in Baltic amber. Because the synchrotron radiation does to X-rays what a laser does to visible light, it gives images with high contrast and a resolution down to a thousandth of a millimeter (0.0004 inch). So Perreau was able to get three-dimensional pictures of the genitalia of the less than 2-millimeter-long (0.1-inch) insect without as much as scratching the surface of the prized amber object. Not that it was a breeze—to do his virtual dissection, Perreau had to tell the computer for each individual detail, sometimes even pixel by pixel (or rather, voxel by voxel, the 3-D equivalent of a pixel), which parts belonged to the genitalia and which did not. "Very much work!" he sighs at the memory of the many days spent behind the keyboard. Still, in the end, what he had on his computer screen was a detailed 3-D rendering—a so-called microtomograph—of a forty-million-year-old, 0.4-millimeter-long (0.016-inch) beetle penis, which he could flip and rotate and inspect from all angles at the click of a mouse button.

The digital ancient aedeagus showed that the beetle belonged to an ancestral species of *Nemadus,* a kind of Northern Hemisphere beetle that scavenges detritus in bird nests. Still, it was clearly an unknown extinct species (which Perreau good-humoredly named *Nemadus microtomographicus*) with a penis shaped differently from all present-day species. Since then, Perreau has brought back to life a few other beetle penises, all distinctly different from the ones sported by their current descendants.

Still, stunning as these amber aedeagi might be, there are simply not enough of them to really reconstruct the history of genital shape and function. What we need is insects of the same family or the same genus represented in amber of a whole range of different ages, so that the changes can be tracked in time. But there are only a few places in the world where good-quality amber comes to the surface, and the amber found in one place is usually of more or less the same age. For most kinds of insect this results in three or four time slices at best. So attempts to use amber insects as an archive of evolutionary changes in genitals are frustrated because, as is so often the case with fossils, they are too few and far between.

A much richer source of information is the insect remains that are often found in peat deposits, where they are preserved in perfect condition because of the absence of oxygen in such environments. Though younger than amber insects—they mostly date back to somewhere within the last half a million years or so—peat fossils come in much larger numbers, especially beetles with their tough, decay-resistant outer armor. The late Russell Coope was a legendary British paleontologist who made those peat beetles his bread and butter, routinely using their preserved genitals to identify the species. In the 1970s he studied a mother lode of 43,000-year-old beetle fragments from the Thames valley in England, consisting of thousands of specimens of almost three hundred different species. And whenever he came across a beetle abdomen, he would extract the genitalia and make perfect microscope mounts of them to compare them against identification keys for present-day species.

The curious thing is that Coope rarely found any sign of evolutionary change in the beetle genitalia, not in the Thames valley specimens or in much older specimens. In a 2004 paper in the *Philosophical Transactions of the Royal Society,* he vents his surprise about this: "[A]lmost all fossil specimens match precisely their modern equivalents. These similarities even extend to the intimate intricacies of their male genitalia, which can be dissected out of compressed abdomens frequently found in the fossil assemblages."

Now this is a bit of a puzzle. We know that closely related species of beetles often have wildly different genitalia, so these must have changed at some point during or after the splitting up of species. And theory tells us that genital evolution should be particularly fast and dynamic. Yet Coope's fossilized beetle penises tell a different story: they speak of stability instead. Of course, there are ways to solve Coope's conundrum. First of all, most of the fossil beetles he studied are just a few tens or hundreds of thousands of years old—perhaps too short for the really big steps in genital evolution. And also, Coope never performed any accurate measurements on the shape of the genital organs. He simply observed that he could use the ancient genitalia to identify their owners as present-day species; small differences may have evolved during the millennia that elapsed since his specimens were alive, but these may have been too small for Coope to notice.

On the other hand, it could also be that Coope uncovered a true aspect of genital evolution, and that these organs do not evolve in a smooth ebb and flow of continuous change, but in fits and starts instead. Whether evolution is usually gradual or erratic is a more general debate in evolutionary biology. There are some, like the late Stephen Jay Gould, the famous Harvard University paleontologist, who maintain that evolution tends to be concentrated in short periods of quick change, interspersed with long periods of stability—a pattern named "punctuated equilibria" by Gould (and "evolution by jerks" by his critics). The alternative is gradualism: smooth, continuous, imperceptible change all the time. This is what Darwin envisaged when he wrote, "We see nothing of these slow changes in progress, until the hand of time has marked the lapse of ages" (and which Gould, with inimitable wit, called "evolution by creeps").

Fortunately, over the past decade or so, scientists have come up with methods to measure whether a particular feature evolves by jerks or by creeps. To do this, you first need an evolutionary tree based on DNA. DNA is made of long strings of four different chemical compounds (the names of which are abbreviated to the familiar A, C, G, and T), and it evolves over time by accumulating mutations—chemical changes in which one or more "letters" are replaced by other "letters," or are lost or multiplied (see Chapter 1). Since these mutations take time to happen, you can use the amount of shared "letters" in the DNA of two species to determine how close they sit together in an evolutionary tree: the fewer differences, the more recently their common ancestor split up. (The techniques are much more sophisticated than that, but basically this is what it boils down to.) So, using the information contained in DNA sequences of a group of species, evolutionary biologists can draw up an evolutionary family tree for them, which can be read from bottom to top as the order and timings of splits of ancestors into descendant species over time.

Mark McPeek of Dartmouth College in Hanover, New Hampshire, employed such DNA-based trees to figure out how reproductive organs evolve in *Enallagma* damselflies. Some forty species of these pretty, dainty, blue-and-black damselflies (also known as bluets) occur all over North America, Europe, and northern Asia, and back in 2005 scientists had already used their DNA to work out the exact evolutionary tree for all of them. What McPeek and his team did was study how the shape of the male damselfly's apparatus of claspers at the end of his abdomen changes along the branches of this family tree.

These tongs-like organs are not strictly speaking genitalia, because of the peculiar way in which damselflies mate. Males hang around ponds waiting for females to arrive from their foraging trips. Once a female is spotted, a male will literally pounce on her in midair and grab her on the neck with those claspers he has. Then he releases a small droplet of sperm from an opening just in front of the claspers and, without releasing the female, bends the tip of his long abdomen forward to transfer this sperm droplet to his actual penis at the base of his abdomen. If the female decides to accept as a mate the male that is holding her, she will, while still dangling beneath him, bend her abdomen all the way forward to make her genitalia meet his and receive his sperm (more on this weird mating system in Chapter 6).

Now, those male claspers differ a lot among species, which McPeek documented by creating 3-D images of them in a CT scanner. Then he used the same sort of software that computer animators use, to work out mathematical formulae that accurately reproduce the three-dimensional shapes of the species-specific claspers. These formulae then gave him a way to measure the differences in shape: the more terms and parameters were different between the formulae for a given pair of species, the more different their claspers had to be. Finally, he compared these measurements against the DNA tree and was surprised to find that there was no relation between the degree of clasper difference between two species and the time since the two species had last had a common ancestor. Old species were no more different than young species. This meant that each species had experienced only a single "jerk" of clasper evolution, rather than a slow process of gradual change. Similar signals of evolution by fits and starts have been found in the genitalia of cactus flies and millipedes.

These results—the fossil evidence as well as these tree-based analyses—demonstrate that, despite rare-male effects and Pomiankowski's sexual tango, there appear to be certain glass walls that limit the kinds and amounts of change that are possible and sometimes cause genital evolution to get stuck in a rut. In that sense, genitals evolve sometimes jerkily, when a new female fad evolves, and sometimes creepily, when it inches closer to one of those glass walls. So we need to adjust our image of genital evolution a little bit: although it does often spiral out of control, rapidly generating extravagant shapes and bizarre forms, it can sometimes also be constrained in an evolutionary straitjacket. In the next section, we'll meet a few such constraints: ways in which genital evolution in certain animals is capped.

Size Does Not Matter

Nobody seems to know exactly where the (much overrepeated) phrase "Size doesn't matter" was first uttered, but its roots lie in the famous sex studies of Masters and Johnson of the 1960s, in which they write, "Another widely accepted 'phallic fallacy' is the concept that the larger the penis the more effective the male as a partner in coital connection." Whether penis length is particularly valued by women or not is a matter of intense and seemingly endless debate in the glossies and scientific journals alike (remember the relation between penis length and vaginal orgasms in Scotland from Chapter 4). But in a zoological sense, average size, rather than extreme size, appears to be at a premium—in humans as well as in other animals.

This has to do with phenomena called isometry and allometry, which are the case when body measurements do or do not scale in proportion, respectively. People with bigger bodies tend to have bigger livers, for example, and for every step increase in weight, you'll find an identical increase in liver weight; somebody who weighs in at a hundred kilos will have a liver twice as heavy as somebody of fifty kilos. So livers are said to have an isometric relationship with body size (*isos* meaning "same" in Greek). But if you take a different organ—say, the brain—you'll find allometry (*allos* meaning "different") instead: heavy, big-bodied people do have bigger brains than smaller people, but the difference in brain size is much smaller than the difference in body size. Such organs, which seem to hover around the same size no matter what happens with the size of the rest of the body, are called negatively allometric. (Positive allometry also exists—the long bones of our arms and legs are disproportionately longer in taller people than in shorter people.)

Now it is a curious rule in nature that genitalia show allometries that are about as negative as they get. Take the stag beetle *Lucanus maculifemoratus*, or *miyama-kuwagata*, as this stunning rusty black beetle is lovingly called in its native Japan (where big beetles like these are routinely kept as pets and department stores have special "insect care" sections). Entomologist Haruki Tatsuta caught forty-seven males—characterized by their gigantic antler-like jaws—in oak forests in Hokkaido and then (perhaps somewhat less lovingly) cut them up into four separate pieces: head with jaws, thorax, abdomen, and aedeagus. Each part he dried in an oven and then weighed. What Tatsuta found was that whereas the dried body of the biggest beetle was ten times as heavy as that of the smallest, his penis was barely one and a half times heavier. Compared with body size, but also

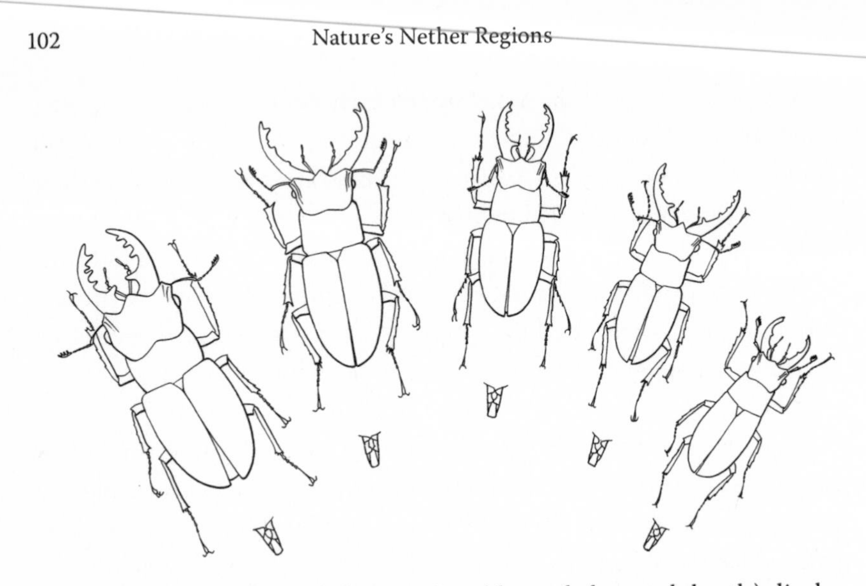

Stag night. Stag beetles and their penises (shown below each beetle) display so-called negative allometry: no matter how large each individual animal is, stag beetle genitals are always nearly the same size.

with the sizes of separate body parts, the beetles' genitalia showed strong negative allometry.

Even in humans, penis size does not vary as much as our preoccupation with the subject would suggest, and it is hardly related to other body size indicators—and certainly not to shoe size, despite the popular myth. In 2002, London-based urologists Jyoti Shah and Nimal Christopher published an article entitled "Can Shoe Size Predict Penile Length?" They somehow persuaded more than a hundred of their patients to have their penises measured and to divulge their shoe sizes to them. Shah and Christopher then graphed both against each other and found no relationship whatsoever—relegating the shoe-penis connection to the realm of fables. However, they measured their patients' members in a flaccid rather than an erect state, arguing that, presumably in view of medical propriety, it was "not feasible" to measure them in their "true physiological length." Point taken, but this could be a problem, because it is known that the longer a man's penis in floppy state, the less it will grow in length when it becomes erect. Nor did the researchers record height or weight or any other body size measure besides shoe size. But in one of the more remarkable citizen science projects, the online questionnaire the Definitive Penis Size

Survey, Shah and Christopher's results are borne out: among more than three thousand men who filled in the questionnaire, their shoe sizes said nothing about the lengths of their erect penises. Stature fared a little better in this respect, but the allometry was still negative: men 20 percent taller on average have erections that are only 10 percent longer.

For as long as biologists have been measuring animals (and that is a long time!), they have found that genitalia, in males as well as in females, tend to be pretty much unaffected by the size of the rest of the body, just as in stag beetles and humans. In a large survey of such animal data, Bill Eberhard, Bernhard Huber, Rafael Lucas Rodriguez, and colleagues found that this was true for virtually all of over 130 species of insects, scorpions, spiders, crustaceans, snails, and mammals. Apparently, males as well as females that are under- or overendowed do not fare well in sexual selection, suggesting that evolution smiles upon a kind of general-purpose, "one-size-fits-all" genitals. This does not seem to be because of any literal mechanical fit between the genitals, since negative allometry turns up in animal species with soft, stretchy genitalia just as much as it does in those with hard, unyielding ones. Instead, Eberhard and his colleagues think the reason may be that male genitalia need to "press all the right buttons" in the female genitals. If his or her genitalia are too large or too small, the relevant knobs do not end up opposite the relevant nerve endings and stretch receptors in the female genitalia. So in sexual selection, individuals that are so well or so poorly endowed in the nether department that they are likely to be a mismatch with many potential partners will not leave many genes in the next generation. Hence, evolution will punish those with extremely small or large genitals, and negative allometry, average-sized genitals, and "one-size-fits-all" will be the result.

Now, the one-size-fits-all rule applies to animals where the male shoves his entire penis up the female's vagina. But there are also species where only the tip of the penis is inserted. When that is the case, the bulk of the penis is no longer constrained by the need to fit in the female's vagina and if, for whatever reason, evolution favors bigger penises, then grow bigger they will. We have seen this with the record-setting barnacles that Darwin discovered (Chapter 3), and in Chapter 8 we will meet slugs that have evolved penises so long that it takes them a whole night to get erect. But even then, limits may apply. Namely, when sexual selection promotes a kind of evolution that the rest of the environment, by way of natural selection, won't allow.

The first example of the environment curbing sexual evolution is in so-called poeciliid fish—the family to which the guppy (*Poecilia reticulata*) belongs, which we came across earlier in this chapter in the context of color and the rare-male effect. Poeciliids are rather exceptional fish. Not only are they among the few kinds of fish in which the males have an actual penis to inseminate the female internally—rather than just dumping their sperm over the eggs as she lays them, which is what most fish do—they also are one of the few exceptions of animals in which penis length is not negatively but positively allometric. In male guppy fish, a body size twice as long translates to a penis—known as a "gonopodium," in fact an anal fin fashioned by evolution to serve as a sperm squirt—not twice but up to four times as long. Poeciliids may be the exception that proves the one-size-fits-all rule, since during mating males do not insert the entire gonopodium into the female. Instead, they swing it forward and either proudly display it in courtship (where females prefer better-endowed males) or—forgoing courtship—take quick stabs at her vagina from a distance, with only the tip penetrating. Interestingly, when you measure the width of the tip, the only part that actually meshes with the female genitalia, you *do* find negative allometry. So, clearly, in poeciliid fish, girth does not matter, but length does.

But, as research in one poeciliid has revealed, there are limits to how long your gonopodium should get. Males of the mosquito fish *Gambusia,* a guppy relative that lives in lakes and ponds in the southern United States, have particularly impressive gonopodia that sometimes reach up to a third of their body lengths, and males have a habit of flashing these to interested females. Thus engrossed, however, they may not notice the predatory sunfish taking advantage of the fact that their guard is down and swooping in to snap up the hapless hormone-heavy poeciliids. Unless, that is, the mosquito fish can make a quick last-instance getaway. It is in these desperate escapes that their big members become a liability. As Brian Langerhans discovered in his work at Washington University in St. Louis, when you're a mosquito fish and you're trying to swim away quickly, a heavy trailing penis is—quite literally—a drag. Langerhans and his colleagues measured the speed of burst swimming in well-endowed males and found that these were much slower than in poorly endowed ones. Not only that, they discovered that the trade-off between sexual selection on penis length (bigger is better) and natural selection (shorter is safer) had left its mark on natural populations: in ponds with predator fish, gonopodia were 10 to 15 percent smaller than in predator-free ponds.

Another example of such a trade-off is found in *Tidarren* spiders. About ten species of these small spiders exist, all living in the Tropics and all characterized by being half-eunuchs. As you will recall, male spiders load their pedipalps with sperm and then usually use both of them to inseminate the female. But *Tidarren* adolescent males, just after their second-to-last molt, do something rather peculiar. While hanging upside down from their web, they use their legs and one of their pedipalps to push the other pedipalp firmly in between the tangled threads of the web. Then they begin circling around the one pedipalp that soon becomes hopelessly tangled in more and more silken threads and, after a couple of revolutions, is no longer able to spin along with the rest of the spider and snaps off. Unfazed, such a self-mutilated male then sucks the amputated pedipalp dry and continues his one-pedipalp life (the amputated palp never grows back) until, after his final molt, he is ready to mate.

Tidarren is one of those spiders in which mating is a once-in-a-lifetime opportunity. Their testes produce a single drop of sperm and then wither away. This single drop of sperm is sucked up in *Tidarren*'s one remaining pedipalp. His one-shot genital apparatus at the ready, he seeks out a female and then copulates with her—only once, for consummation is consumption: the female invariably has him for lunch even before copulation is over. As she begins nibbling away at his body, he rests secure in the knowledge that meanwhile his pedipalp—still firmly attached to her epigyne—is busily pumping his life's supply of sperm into her.

The self-semi-emasculation is the outcome of a curious set of evolutionary circumstances. In *Tidarren,* females have evolved large body size—presumably because bigger females outcompete small ones in terms of egg output—while at the same time males have evolved small bodies. And although the cause for the miniature males is still unclear, the result of this evolution in opposite directions is that *Tidarren* males weigh less

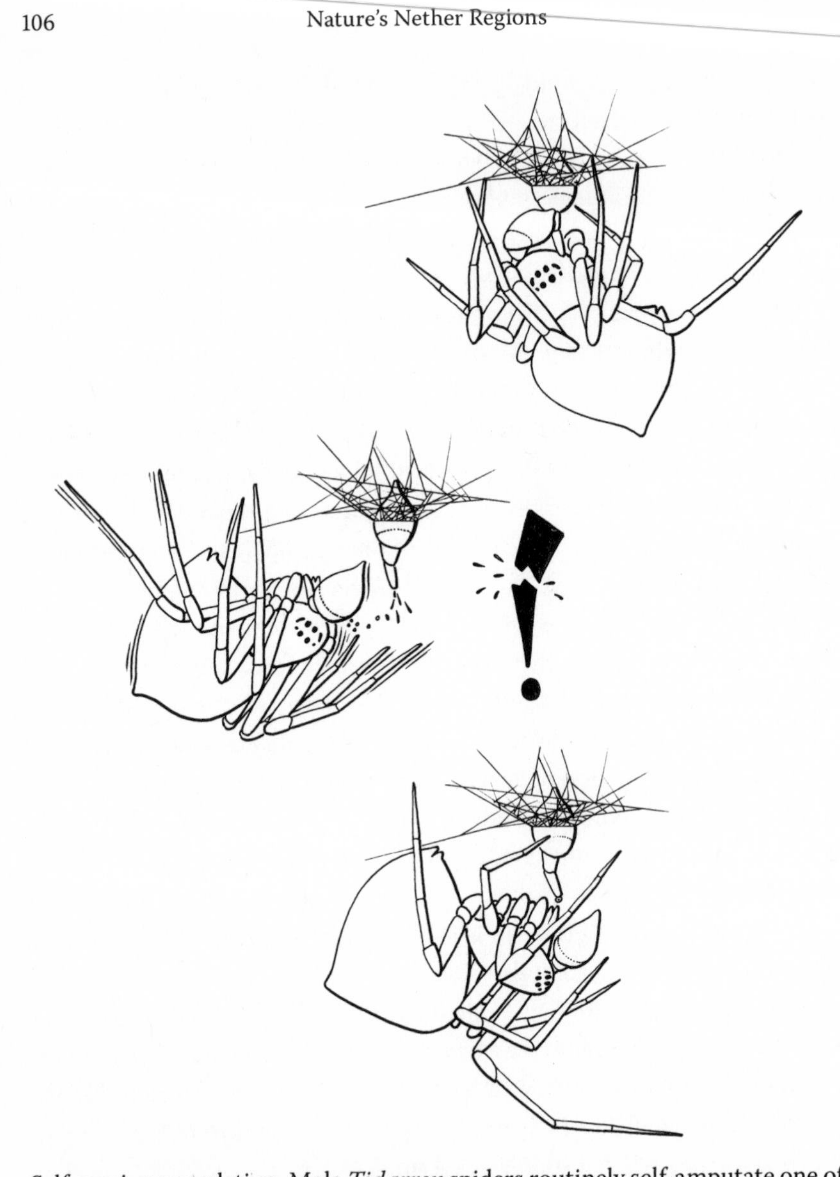

Self-semi-emasculation. Male *Tidarren* spiders routinely self-amputate one of their pedipalps, the better to move about. They do this by entangling one pedipalp in the silk of their web and then turning around and around until it snaps off. Then they suck the amputated pedipalp dry.

....

than 1 percent of a female's weight! To still be able to fertilize a female, they need very, very large pedipalps—each weighs over one-tenth of the male's body weight. The problem is, with two such hefty pedipalps hanging in front of your head, it's almost impossible to move around. Hence, the evolution of the half-eunuch strategy.

That natural selection for improved mobility is indeed likely to have been the impetus for this bizarre behavior was proven in a 2004 study by Margarita Ramos and colleagues at Tulane University in New Orleans. They placed male *Tidarren sisyphoides* spiders on a tightly stretched silken thread of a female spider, either just before the males had self-emasculated or just after. They found that males still in possession of both pedipalps scooted along the thread at a speed of less than 3 centimeters (1 inch) per second, whereas half-emasculated males did so at more than 4 centimeters (1.5 inch) per second. Not only that, their endurance was also severely affected by the presence of an extra pedipalp. The researchers used a small paintbrush to chase a male around a piece of paper. If the male still had both pedipalps, he would collapse from exhaustion after about twenty minutes, whereas males that had self-amputated one pedipalp persevered for more than half an hour before giving up.

The bizarre self-inflicted genital mutilation of male *Tidarren* spiders clearly means that its pedipalp has evolved to a size that is just about as large as the demands of a functioning spider will allow. Any larger and it would be too heavy and unwieldy for a *Tidarren* male to drag himself into his nuptial-cum-deathbed—that is, a female's arms.

Pedipalp-amputated spiders, gonopodium-encumbered poeciliid fish, and all animals that have one-size-fits-all genitalia demonstrate the limits to what sexual selection by cryptic female choice can do. And perhaps the hitting of these limits is what we see when we chart those jerky evolutionary pathways of genitalia in fossil beetles and throughout the family trees of damselflies and other animals. Perhaps once such a limit is hit, sexual selection is lame for a while and has to wait for a new mutation to appear so that genital evolution can dart off in a different direction.

Does this mean that our picture of genital evolution is now complete? With sexual selection in the driver's seat, steering the evolution of animals' genitals along a winding road curbed by the limits of natural selection, it may seem that we have a satisfactory outline and that the rest is just details. But when all seems crisp and clear and yet you're only halfway through a book, a plot twist lies around the corner. As we shall see in the next few chapters, genital evolution is not only about male courtship, female control, and their limitations. Prepare for sexual persuasion to turn nasty.

Chapter 6

Bateman Returns

Entomologist Jonathan Waage of Brown University did for the penis what the Swiss army did for the penknife. In a two-page article that appeared in the journal *Science* of March 2, 1979, Waage revealed how a male ebony jewelwing damselfly (*Calopteryx maculata*) will use his penis to scoop out any sperm of previous males from the female's genitals before using it to deposit his own. The penis as a sperm-scooper. The compact publication is seen by today's researchers as the foundation stone of the whole field of genital evolution, and it makes its appearance in many a biology textbook. But at the time, it created little more than a tiny ripple.

"I do not recall more than a few articles that showed up here and there in newspapers. No interviews," says Waage. Still, the news did manage to make it across the Atlantic. I was only thirteen at the time, but I do remember reading about it, probably in *Kijk,* the Dutch high school science and technology monthly that I used to devour the moment it landed in our mailbox.

What I read, thumbing the pages of *Kijk* excitedly, was that Waage had tried to figure out what happens in a female damselfly's sperm storage organs when she mates with more than one male in short succession (which damselflies usually do). Are the sperm of the second male simply mixed with the first? To tackle this question, he began by catching female jewelwings—easily recognized by the white spot on their wings—at the bank of the Palmer River, a small stream just across the border in Massachusetts. Each time he caught a female, he carefully tied a length of nylon fishing line to her abdomen and, the female thus tethered, gently moved her into the territory of a male, which promptly elicited a copulation.

You may recall from the previous chapter that damselflies mate in an unusually cumbersome and acrobatic way. This is because damselflies and dragonflies are the only insects in which the male's penis (at the base of the long, wispy abdomen) is not attached to the testes (sitting at the tip)—a similarly disconnected affair as in spiders, which are not insects, of course. To deal with this sexual handicap, a male, after grabbing a female in the neck with the pinchers at the end of his abdomen, needs to transfer the droplet of sperm that his testes produce to a temporary sperm container next to the penis. He does this by doubling over the entire abdomen, basically to inseminate himself. Then the female throws *her* abdomen forward to meet his penis, thus forming the romantic heart-shaped "mating wheel." In this position they fly or hang around for a variable length of time—a couple of minutes in the ebony jewelwing, but up to several hours in other species.

In some cases, after a male had released her, Waage let a second male step in and mate with his female-on-a-leash; in some cases not. Either way, when the last (or only) copulation was over, he would invariably end the affair in a rather unceremonious way: by dunking the female in ethanol. Back in the lab, after dissecting the sperm storage organs of the dead females, he found that, whether they had mated once or twice, they always carried the same amount of sperm. Puzzled, he went back to the stream bank and repeated his first experiments, this time always allowing the females to mate with two different males, but killing the females before completion of the final coitus. Again, he dissected the females' abdomens, and found that most of the females held no sperm at all or only very little—which meant that, somehow, they had lost all the sperm after their first copulation.

Once again, Waage returned to the Palmer River and collected pairs of the damselflies at different stages throughout the mating, which he killed *in copula* and then carefully dissected under his microscope in an attempt to figure out what was going on inside a pair's genital organs. This finally clinched it for him: in couples killed early in the copulation, Waage saw that the male still held all his sperm in his sperm store, but his penis was firmly inserted in the female's vagina, and large masses of sperm (presumably from the female's ex-boyfriend) clung to the backward-pointing bristles on the penis and the scoop-like horns on either side. In couples arrested in the very final stage, however, the male's sperm store was empty, but the female's sperm storage organs were once again filled (with her incumbent lover's sperm). So it turned out that to a male *Calopteryx maculata,* copulation is about removing rival sperm just as much as about depositing his own. This also explained why the shape of the damselfly mating wheel passes through two stages. In the first part of copulation, the male's abdomen is held in a concave shape and is seen to undulate constantly, but in the last few seconds it changes to a convex shape. What Waage's study had shown was that the first stage is devoted to vigorous

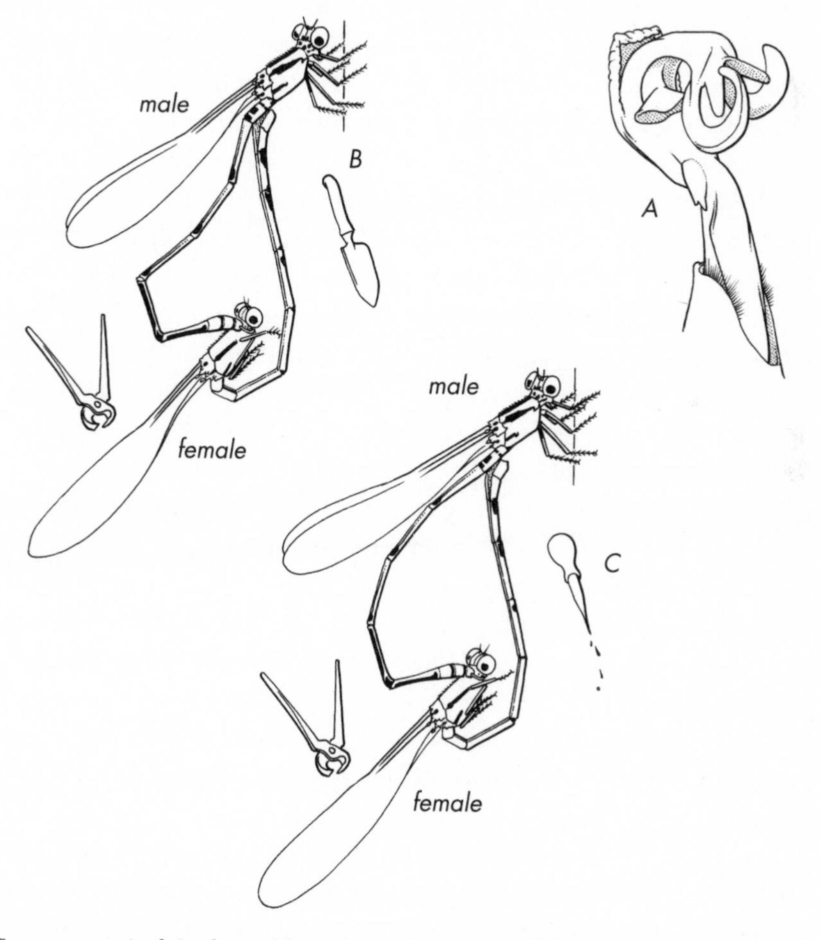

Scoop your rival. In damselflies, the male grabs the female in her neck and she then flips her vagina forward to meet his genitals. In some species, the male uses the scoop on his penis (A) to remove sperm from previous males from the female's vagina (B) before injecting his own sperm (C).

....

cleaning out of any previous sperm from the female's genitals, and the second stage to filling her up with the male's own.

Waage's article in *Science* was a seminal publication (pardon the pun). It kick-started more research on damselfly sex, and a worldwide community of researchers has since confirmed that what happens in the ebony jewelwing is common practice in most other damselflies and probably also in their bulkier relatives, the dragonflies. It is now known how sense organs on the penis register the presence of foreign sperm in the female's genitals and how the flaps, hooks, and microscopic spines and teeth on the penis trap the clump of sperm while the muscles of the penis move it up and down to pump, scoop, and scour out sperm of a predecessor.

More important, Waage's paper, published six years before Eberhard's first book, marked the start of the current wave of serious, wide-ranging research on genitalia in all kinds of animals. Waage had shown, in spectacular fashion, that damselflies adhere to a very literal interpretation of what Geoff Parker (of dung fly fame) had termed "sperm competition." Parker, in a famous 1970 article, had suggested that a male insect, rather than just persuading a female to choose his sperm over that of his competitors, may up his chances by targeting that competing sperm directly. And indeed, Waage's damselflies did exactly that, using their penises as true shovels to purge their mates' vaginas of any rival spermatozoa. Hence, we might expect that the wave of ensuing genital research that began after Waage's publication was driven by the notion that scuppering the reproductive efforts of one's rivals might be a very important force in the evolution of male genital organs. And that the road lay open to examine all genital form in this light.

But this did not happen—or at least, not immediately. Waage himself continued working on damselfly sex throughout the 1980s, but eventually moved into administrative positions, spending less and less time on his scientific work. And although he fathered a band of damselfly sexologists who followed in his footsteps, in the meantime Bill Eberhard's first book was published, which stole the spotlight. The growing community of animal genital researchers turned to cryptic female choice, overlooking the importance of sperm competition. And yet, if we take a closer look at the way male and female genitalia interact in different species of damselflies, we get a peek into a world that does not seem to be ruled entirely by cryptic female choice, but rather by a kind of evolutionary loggerheads over what he wants and what she wants.

A couple of years after his *Science* article, Waage turned to spreadwing damselflies, so named because they do not properly fold their four wings when they alight, but leave them hanging in a rather disheveled fashion. The females of one spreadwing species, *Lestes vigilax,* have a sperm storage organ that is very spacious—it can accommodate more than one male's ejaculate, and its large size makes it more difficult for the male to scrape out any predecessors' sperm. Not to be outdone, males still manage to keep their sperm in the game by using their penis to pack any existing sperm into out-of-the-way corners of the female's sperm storage organ, where it is not likely to be used for fertilizing any eggs. If the score was 0 for the girls and 1 for the boys in the first species that Waage studied, in *Lestes vigilax* it's girls 1, boys 2.

Among *Calopteryx xanthostoma,* a species related to the one that Waage did his pioneering work on, the deeper recesses of the female's sperm storage organ are so narrow that the flaps on the male's penis cannot get in. Given his penis's bulk, he has to accept that some rival sperm will remain unreachable to his sperm-scooper and that many eggs laid by her will be fertilized not with his, but with other males' sperm hiding deep inside her. Girls against the boys: 2-1.

The same would be true for yet another species, *Calopteryx haemorrhoidalis*—which, like *C. xanthostoma,* also has penis horns that are not slim enough to penetrate the deeper female genitalia—were it not for the following clever trick, revealed by Mexican entomologist Alex Córdoba-Aguilar. All damselflies have two plates in the walls of the vagina. The sense organs embedded in these plates feel when an egg passes through the vagina and send a signal to the spermatheca; this elicits the release of a droplet of sperm to fertilize the egg with. What has happened in *Calopteryx haemorrhoidalis* is that this female fertilization system has been hijacked by the male. The shape of the penis is such that, each time it pushes into the female vagina, it buckles the vaginal plates in the same way that a passing egg would. The sense organs in the vaginal plates, duped into "thinking" that an egg is being laid, send a signal to the spermatheca to release previous trysts' sperm, which drips out of the spermatheca and is removed by the male with his penis. There is no egg—but the female physiology is tricked by the male's egg-shaped penis, and the male wins the game after all.

To show that the *C. haemorrhoidalis* penis indeed has a shape that persuades a female to empty her sperm stores, Córdoba-Aguilar has done experiments in which he tied down females of other species and, holding a detached *C. haemorrhoidalis* penis in the tips of a pair of tweezers, stimulated their vaginas in the same way that a copulating *C. haemorrhoidalis* male would. And, lo and behold, these other species also helplessly released large quantities of the sperm held in their spermathecae.

In yet other species of damselfly the female has got the upper hand again. A female of *Paraphlebia quinta* from forested hill streams in Central America will often interrupt a copulation after a male has finished scraping but before he has begun depositing his own sperm—thereby losing her sperm stock, but denying the male the opportunity to replace it with his own. And even if she allows the male to ejaculate into her, after mating she often ejects a small droplet from her vagina: the familiar, if unsavory, act of sperm dumping.

What this little odonatological foray shows all too clearly is that damselflies do not seem to fit with the picture of penile persuasion by internal courtship that I have painted in the previous chapters. Yes, the males have penises that are shaped differently in each species, and, yes, they do pump them up and down in the vagina of the female, who may or may not decide to remain in mating wheel position until the ejaculatory culmination. But against this familiar backdrop of internal courtship and cryptic female choice, the plot thickens into a genital drama of rivalry, male-female conflict, and deceit. Males use their penises to manipulate other males' luck in love, as well as females' sexual autonomy. And damsels in distress evolve ways to limit the success of males in getting their way with them. Experts call it "sexually antagonistic coevolution," which is a mouthful but reveals quite accurately what goes on. It is sexually antagonistic in the sense that what evolves in males is not always good for the females, and vice versa. And it involves coevolution, meaning that both sides, though antagonizing the other, evolve hand in hand: an evolutionary step taken by females is sooner or later countered by an evolutionary response from across the sexual divide.

Sexually antagonistic coevolution is, in fact, unavoidable. Cryptic female choice sets up a conflict of interest during each and every copulation. For the female involved, it is best to maintain control over the fate of any male's sperm. She needs to weigh her options: if it is an extremely handsome male, then passing his attractive genes on to many of her sons would be a good strategy. If he is a lesser male, it may be good to use his sperm only for a small proportion of her offspring, if at all. In any case, not having all her eggs fertilized by a single male will usually be a good thing. By creating genetic diversity among her clutch she reduces the risk that all of them will succumb to a disease or carry some genetic defect: she'd better not put all her eggs in one genetic basket. So, in evolution, female tubing and wiring that maximize her ability to make these (unconscious) choices will be favored. But for the male, these calculating intentions of his mate are bad news. For him, it would be best if the female were to fertilize as many eggs as possible with his sperm, and evolution will allow any ways in which he can coax, coerce, or con her into doing so.

So even if the evolution of sex in a species may start off in a pleasantly cooperative way—males show females what they have on offer, females make their choice, the unsuccessful males try their luck somewhere else, no hard feelings—it can quickly escalate to an evolutionary concatenation of tricks to get the upper hand in the sexual conflict. As Richard Dawkins and John Krebs have written, "[S]words get sharper, so shields get thicker. So swords get sharper still." Such a conflict, remember, is inherent in Bateman's principle: males gain more reproductive success the more sperm they manage to get accepted by females, whereas females gain reproductive success by choosing their males prudently. In other words, for males sex is more about quantity; for females it is more about quality. Hence those sperm-scoopers, unfathomable female sperm stores, and egg-mimicking penises in damselflies.

As you will realize, our story now enters uncharted and treacherous territory. Here be monsters. In the next few chapters we will observe the dubious role genitals may play in these evolutionary arms races throughout the animal kingdom. But we should always remember that casting these evolutionary escalations in human terms can be misleading. Tempting and illustrative as it may be to speak of the "battle of the sexes," "evolutionary arms races," or "tug-of-war," I will try to avoid this as much as possible. Not only because it taints the sex acts of these innocent animals with the uglier kinds of human aggression, but also because the comparison is faulty. In warring factions, the two groups that are in conflict with each other will always remain separate and every point scored by a group member will benefit his clan only. Males and females are not to be confused with such separate clans, for the simple reason that they spawn their own enemies: males also get daughters and females also get sons. So it may actually be beneficial for a female to mate with a male that manipulates or overrides her cryptic female choice, if her sons inherit the same ability and will play that trick upon *their* mates in the next generation. Once again, in sex, all is not as it seems. . . .

Sperm Bank Robbers

One cannot help but wonder how the financial office of the State University of New York reacted when it received reimbursement claims from the psychology department for latex dildos and vaginas. But then again, perhaps it was already used to the unusual research needs of biopsychologist Gordon Gallup on the Albany campus. Famous for his early work on the ability of chimpanzees to recognize themselves in a mirror, he has since focused his attentions on the evolution of human sexuality, and has not shirked from producing unorthodox articles with such titles as "Does Semen Contain Antidepressant Properties?," "The Unique Impact of Menstruation on the Female Voice," and "On the Origin of Descended Scrotal Testicles."

In 2002, he and his student Rebecca Burch decided to take on a new project: figuring out whether the same sperm-scooping trick so prevalent in damselflies was also at work in our own species. Aside from the matter of how such scooping should work mechanically, this may seem a rather strange idea at first glance. After all, unlike in damselflies, human sperm cells survive for a maximum of two to four days inside a woman's body, so how likely is it that men face true sperm competition? How often would their penises meet live, foreign sperm inside the vagina of their mate?

Well, perhaps more often than you'd think, as Gordon and Burch found out. They interviewed almost five hundred female New York college students and learned that 12 percent say they have had, at any point in their lives, sex with more than one man in succession within a twenty-four-hour period. And one in twelve reports having engaged in group sex with multiple men at the same time. Sperm competition may be a thing to be reckoned with, after all. On the grimmer side, sex violence statistics from several countries show that about one-fourth of all rape cases involve multiple perpetrators. If similar figures would have applied in our evolutionary past, sperm-scooping ability may have evolved in humans as it has in damselflies.

But how could it work? This is where the products of California Exotic Novelties, Inc., come in. In an article published in the journal *Evolution and Human Behavior*, Gallup and Burch and their team report how they used the latex genitals to test whether the rim behind the glans on the human penis might be effective at displacing previous males' sperm from a vagina. They went about it with gusto. Testing various recipes for "artificial semen" by mixing water and cornstarch, they reached a texture and consistency that, as they write, "three sexually experienced males" deemed similar in texture to human semen. They placed an ejaculate-sized amount of the mock sperm deep inside the artificial vagina, and then inserted their dildos. The wall of the vagina was helpfully transparent, which made the fate of the cornstarch semen clearly visible. It worked! When the dildo went down, the semen in the vagina flowed via the space around the frenulum (the stringy bit of tissue bridging the glans and the shaft) and got trapped behind the rim of the glans. When the researchers pulled the dildo up again, this piston-like action lifted much of the semen up and out of the vagina. Thus, some 90 percent of the semen was scooped out with every stroke—at least when "lifelike" dildos were used. They also tried a smooth vibrator (switched off) without a glans, but this one hardly removed any semen.

Could this mean that the shape of the human penis has partly evolved to tackle sperm competition? It's an intriguing thought, but I am skeptical. Our closest relative, the chimpanzee, engages in much more "group sex"—where female chimps associate with groups of males and mate with several males in succession—yet the chimp penis is smooth, dagger-like, with no scoop-like glans whatsoever.

That is not to say that sperm scooping does not occur outside the realm of damselflies and dragonflies. In fact, it is probably quite widespread. The male flour beetle (*Tribolium castaneum*), for example, has a furrow on his penis studded with bristles, which proves very effective at brushing out other males' sperm from females. So effective, in fact, that it can backfire, as a Belgian-British team of researchers found out. When a male, his rival's sperm still sticking to his penis, dismounts from one female, only to run into another, he may actually fertilize this new female not only with his own sperm but also with the sperm he has just removed from his previous mate: paternity by proxy.

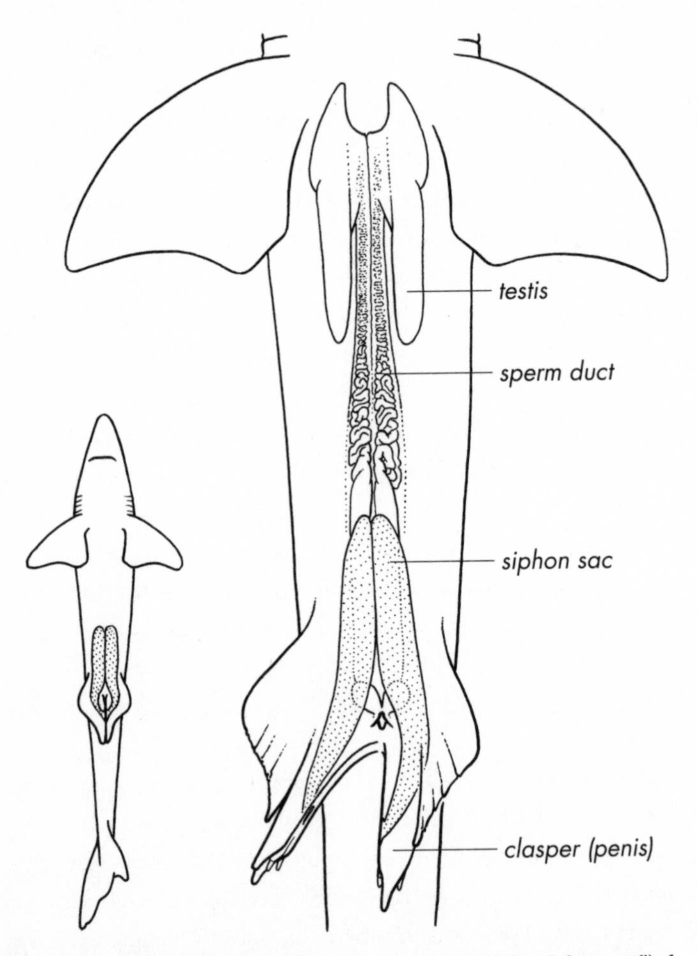

Douche bag. Male sharks, which carry two penises (or "claspers"), have a so-called siphon sac, with which they can squirt seawater into the vagina of their mate, perhaps to rinse out previous males' sperm. The siphon sac as well as the sperm duct both exit through a groove in the claspers.

....

And some animals may flush, rather than scoop. Male sharks are thought to give their mates a kind of vaginal douche before injecting their own sperm. I say "are thought to" because depressingly little is certain about sex in these iconic but little-understood fish (until the 1970s, shark mating had been observed only twice!). What is known is that sharks are blessed with two penises—actually the modified tips of their pelvic fins—each ending in a complex system of movable flaps, prongs, and levers (which is why the penises are normally called claspers, as biologists originally thought they were used to hold the female). During mating—which seems a rather violent affair, involving the male biting the female in the head or pectoral fin, wrapping his body around her, and pinning her to the seafloor—a male will push one of his two penises into her vagina and use the set of flaps to open her oviduct. Along each penis runs a groove that is connected to the testis as well as to the so-called siphon sac, a muscular bladder in the belly of the fish that can suck up a surprisingly large amount of seawater; in the dusky smooth hound, a shark from the western Atlantic, this large reservoir occupies a third of the body length. The siphon sac seems to provide the hydraulics to jettison the sperm into the female, and some shark researchers claim it is also used for swilling out the vagina. Whether this is true is still a matter of contention, but since groups of males often harass single females, each mating with her in turn, such sperm flushing would seem a sensible thing to do.

Crickets also flush. Except that they don't use water but their own semen. The pretty green Japanese tree cricket *Truljalia hibinonis* has a huge (relative to its body length) member that it can push all the way up the female's spermatheca. When it then ejaculates, its sticky mass of sperm will force backward any sperm that is already in there, pushing it out of the female's vagina and onto the penis shaft of the male—who then bends over and proceeds to nibble away at the freshly removed rival sperm as a postcoital snack! And in another cricket-like insect, the European katydid *Metaplastes ornatus,* sperm flushing is a part of sex in which male and female actually collaborate. The male inserts his genitalia, moves them in and out for a bit, and then withdraws. Since his genitalia are jagged and crenellated, pulling them out of the female brings forth a large portion of the female reproductive system, momentarily exposed inside out. What then happens is rather remarkable: the female doubles over and licks out the inside of her own genitalia, eating up any sperm packages of previous mates that still remain in there. This ritual is repeated several times before the male finally deposits his own sperm.

This last example may seem a little puzzling: why would a female join in on a male's attempts to empty her sperm stores? Again, we have to remember that a female has a vested interest in whatever the male does. A male that is able to persuade her to give up her stored sperm reserves probably has qualities that she'd do worse than to pass on to her sons. So, in some species, the whole sperm-scooping business has evolved to become incorporated into the mating ritual and, in a way, has become part and parcel of cryptic female choice—merging imperceptibly with sperm dumping.

Rather than converging on cooperation, another outcome might be an evolutionary race over which sex is in charge of a female's stored sperm. One obvious way in which such a race could be run is by the male and female alternately taking an evolutionary step to take control, like the variety of measures and countermeasures we have seen in damselflies. The female may evolve a somewhat deeper spermatheca, pulling her sperm reservoir out of reach of the male. Sooner or later, a mutation on the male side may then appear that either gives him a penis extension or some other way to regain his sperm-scooping ability. This may then be followed by yet a further deepening on the female side, and so on. Since there is no way back, such evolutionary races might eventually spiral out of control, leading to ridiculously long penises and ditto spermathecae.

One kind of animal whose innards seem to have been seized by such unstoppable evolutionary convulsions is the rove beetle. Chances are that you have managed to live your life without rove beetles ever having claimed any of your brain waves. Let me remedy that lapse, for rove beetles, with their sleek, somber bodies, sit at the high table of insect biodiversity. They come in all colors, sizes, textures, and walks of life, but your standard rove beetle is a few millimeters (about one-tenth of an inch) long, dark, and has its wings tucked neatly underneath shortened wing covers that leave much of the long abdomen uncovered—a bit like a tiny earwig, which they resemble but are not related to. They move fast, and when alarmed stick up their abdomens in threat, though there is no sting in the tail. To biologists (who prefer to call them Staphylinidae), their nastiest habit is their unidentifiability. Some fifty thousand different species are known, and counting (fast), with experts estimating that in reality at least four times as many exist, most living unobtrusive lives in the soil. In fact, it's likely that there's a rove beetle underfoot with every step you take in a forest or meadow.

Though most rove beetles are predators, quietly catching and digesting even less glamorous creatures like springtails, mites, and insect larvae, some lead rather more bizarre lives. *Aleochara* is one of them; it belongs to that select group of insects with a so-called parasitoid lifestyle. Reminiscent of the movie *Alien,* the habit of a parasitoid (which means "parasite-like predator") to grow inside the body of another animal—to eat it from the inside out while the "host" remains alive and active until the parasitoid ruptures the skin and emerges—strikes most people as gruesome rather than "fascinating," the term preferred by *Aleochara* aficionados like Klaus Peschke and Claudia Gack of the University of Freiburg in Germany.

Peschke and Gack have been studying two European species of *Aleochara: A. curtula* and *A. tristis.* Both are parasitoids of flies—their larvae seek out the pupae of blowflies in dung (*A. tristis*) or animal carcasses (*A. curtula*). (If you're going to be gory, then why not go all out?) They nibble a hole in the shell of the pupa, squeeze in, and then start sucking away at the developing fly inside. In the end, the rove beetle larva is ready to pupate into a mature beetle, leaving only the empty husk of the fly pupa behind.

But Gack and Peschke have risen to biological fame not for unveiling the *Aleochara* life cycle, but for the beetles' sex game of who draws the shortest straw. Let's start with *Aleochara curtula,* the species that frequents cadavers. Like all *Aleochara,* they mate in a rather unusual "head over heels" fashion. The male crouches behind the female and then, like a contortionist, flips his long abdomen over his head to couple with the female in front of him. Once securely fastened inside her vagina (and forced to pace along with her in a rather awkward manner if she decides to move about while mating), he unfolds from his penis the internal sac that, as in Jeannel's cave beetles of Chapter 2, inflates only upon "erection." Although the internal sac extends the effective length of the penis, this is still nowhere near long enough to reach into the female's spermatheca, which lies behind a sort of valve, at the end of a long, narrow tube. But the male has another trick or two to play. The first of these is his so-called flagellum, a long, thin rod affixed to the internal sac that, when the sac inflates, gets pushed into the female's narrow spermathecal tube. All along the length of this rod runs a furrow, creating an open space that blazes a trail for the next trick to appear from the male's penile sleeve: a self-inflatable spermatophore.

You remember spermatophores, right? They entered this book in a painful way by ejecting themselves into diners' gums in Chapter 1 after the consumption of freshly deceased male squid. Like squid spermatophores, the ones produced by *Aleochara* also seem to have a life of their own. As the male begins to withdraw his genitalia, a tube starts to grow from the far end of the pear-shaped spermatophore he has left inside the vagina. Fueled by osmotic pressure only, this tube quickly fills the space left by the flagellum, traveling up the tunnel leading to the spermatheca, and forces through the valve at the end. To complicate things, inside this traveling spermatophore tube, an "inner tube" filled with sperm begins to grow, quickly catching up with the leading tip of the first tube and eventually ballooning to fill the entire space of the spermatheca, meanwhile—and this is the crucial part—forcing out, via the valve, any sperm of previous males lingering inside. Finally, a sharp tooth inside the female's spermatheca tears through the spermatophore and the semen is released for further use.

"Why easy if you can also do it complicatedly?" asks Gack. This complex system of getting your sperm inside a female seems to have evolved via several evolutionary steps. The female, by hiding it behind a long tunnel and a valve, has "tried" to keep the male from putting his sperm directly into her long-term storage site. Meanwhile, the male has evolved a succession of penis extensions: the probe-like flagellum on his internal sac and an autonomous sperm package that finds its own way up the female's labyrinth and manages to push out previous males' sperm. Still, it's important to note that, in the end, whether a male's sperm is going to be released in the spermatheca is still under female control, because it is she that has to flex her inner muscles to make that large tooth inside her spermatheca rupture the bag of semen. So, again, the race is more like a game with rules set by the female.

Bizarre as all this may seem, the strangeness of *Aleochara* does not end there. I did not yet mention the length of the flagellum on the internal sac of *Aleochara curtula:* it is about 1.5 millimeters (0.07 inch). For a beetle of slightly under 10 millimeters (0.4 inch) in body length, that's sizable but not extraordinary. In *Aleochara tristis* (the one found in dung), however, the evolutionary contest between the length of the flagellum and the duct to the spermatheca seems to have run a bit out of control. This beetle is about 6 millimeters (0.25 inch) long, but its flagellum is a whopping 16 millimeters (0.6 inch) and is held rolled up like a spring inside the male's penis when not in use. And so is the equally long duct of the spermatheca inside the female.

Carrying on with such a formidable legacy of millennia of sexual selection is not easy for these beetles, as Gack and Peschke found out when they observed the beetles mate. Hard as it may be for the male to thread all sixteen millimeters of his flagellum up the female's narrow duct, even harder is it afterward to stow it back in a neat, ready-for-use coil inside the penis. If, after the deed, the male would simply pull his flagellum out of the female, its springiness would immediately make it tangle up hopelessly, effectively emasculating him for good. So how do they do that? Gack and Peschke wondered. And as the beetles performed their dismounting ritual under their microscopes, the researchers' amazement mounted concomitantly.

As it turns out, the males literally sling their long flagellum over their shoulder. A male first pulls his flagellum a little bit out of the female's vagina. Then, his abdomen stuck into the air, he tilts his chest a little bit sideways, creating a notch between thorax and wing covers, and clamps the flagellum loosely in there. Thus secured, he gently continues to pull the flagellum out of the female by straightening his abdomen, all the while holding the flagellum taut to prevent any knots or tangles, and slowly begins feeding it back into the space inside his penis. As the flagellum is halfway in, he turns around, pulls the last bit of it out of the female, and, still keeping it tight for as long as possible, continues to push it back into the penis like one of those spring-loaded measuring tapes.

Aleochara tristis is not alone in the long-flagellum business. Taxonomists have found similarly lengthy flagella in the male genitals of many other beetle and bug species, which probably function in a similar way—most just haven't been studied in detail yet. It remains to be seen whether all are the result of sexually antagonistic coevolution. The genitals of other,

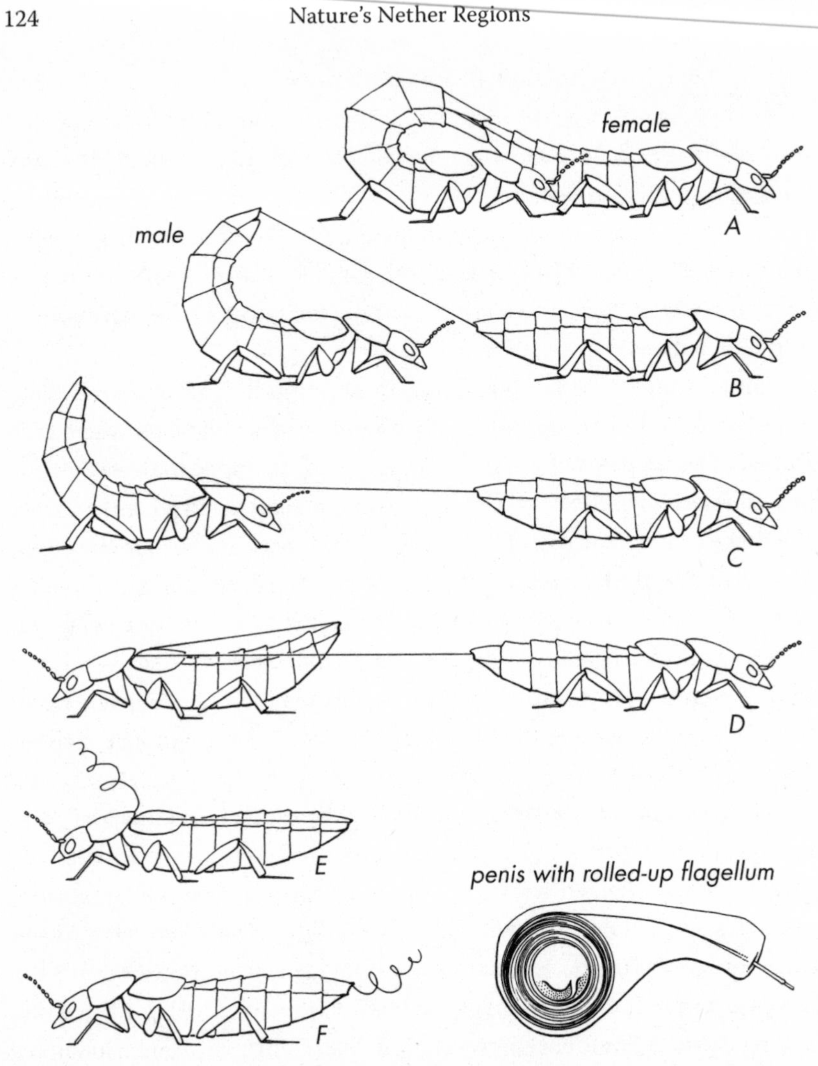

A tape measure up your back end. The male rove beetle *Aleochara tristis* carries a so-called flagellum in his penis, which he keeps tightly rolled up when not in use (bottom right). When it's in use, he threads it up the female's vagina during mating (A), and afterward has to pull it out of her (B), halfway clamping it under his shoulder (C–E), and then reel it in again (F).

more familiar animals, however, have such male-female contests written all over; in the next section, we shall learn why the duck family comprises the bird species with the world's longest penises and deepest vaginas. Brace yourself for some unsavory sex.

Not for the Fainthearted

This story begins with two famous dead drakes. One is an Argentine lake duck (*Oxyura vittata*), shot in Argentina's Córdoba province and stored in the collection of the University of Alaska's museum. We'll get back to that one in a short while. The other is specimen NMR 9989-00232 of the Rotterdam Natural History Museum, a male mallard (*Anas platyrhynchos*) that ended its life with a bang—in more than one sense—on June 5, 1995.

On that day, at five minutes to six in the afternoon, museum curator Kees Moeliker looked up from his work as his ears registered the familiar thud of a bird hitting the museum's glass facade. Despite the bird silhouettes painted on the new building's exterior, birds kept flying themselves to death at regular intervals, and Moeliker had resigned himself to bagging the deceased pigeons, blackbirds, and waterfowl for the museum collection. But as he descended the stairs to see what large bird had this time bequeathed itself to the museum's holdings, he was in for a surprise. At the foot of the glass wall lay a very dead male mallard duck, facedown on the sand. Next to it stood another, very much alive drake in a state of great sexual arousal, which, as Moeliker watched from behind the glass, proceeded to mount his dead companion and start copulating with it with great abandon.

Rather perplexed, Moeliker sat down on a guard's stool and, always the observant biologist, pulled out his notebook and began recording what he saw. For one and a half hours, the live male continued mounting and dismounting and repeatedly copulating with his dead conspecific. By 7:10 p.m., having covered a full page of notes, Moeliker decided he had had enough of his first observation of homosexual necrophilia, shooed off the still testosterone-heavy live male, stuffed the brutalized corpse of his dead lover in the museum freezer, and went home for dinner.

Only six years later, after having been repeatedly nudged by his colleagues, Moeliker got around to publishing his observation in the museum's scientific journal. In his paper "The First Case of Homosexual Necrophilia in the Mallard *Anas platyrhynchos* (Aves: Anatidae)," Moeliker tried to make sense of his observation. As an ornithologist with a special interest in the uglier aspects of ducks' lives, he knew that besides monogamous pairing, male ducks also engage, often in groups, in what can only be described as gang rape. Not seldom such sexual frenzy escalates in numerous males, quacking loudly, giving chase to a single female. Such so-called "rape-intent flights" usually end when the drakes eventually corner the female in some park pond, where, much to the consternation of human duck-feeding bystanders, she then almost (or sometimes actually) drowns under their repeated "attentions."

Apparently, mate choice is not something that interferes very much with the sex drive of such males: for a drake who's single and aroused, any duck will do, be it drake or dame, dead or alive. Bruce Bagemihl, in *Biological Exuberance,* his amazing book on the homosexual animal, writes that drakes often copulate with other males or with dead females. Apparently, it was only a matter of time before the opportunity presented itself to a drake to combine both inclinations—in this case right in front of an interested ornithologist. Somehow, Moeliker's paper caught the attention of the committee for the Ig Nobel Prize, an annual fun-filled scientific parody event at Harvard University, where in 2003 he was awarded the Ig Nobel Prize for biology. As a commemoration, each year on the anniversary of its unglorious death, specimen NMR 9989-00232 is honored at the Rotterdam Museum with Dead Duck Day—capped off with a communal dinner party (Peking duck) in a local Chinese restaurant.

The habit of forced copulation, which is rife among males in the duck family, and which led to the Rotterdam mallard's death and subsequent defilement, is also responsible for the most eye-catching aspect of the second dead duck in this story. The journal *Nature* of September 13, 2001, ran a one-column article by duck researcher Kevin McCracken of the University of Alaska Museum of the North and his colleagues. Despite the paper's brevity, the accompanying Figure 1 was hard to miss. It showed a dead male Argentine lake duck (*Oxyura vittata*) suspended from its wings against a whitewashed wall, with a drooping 42.5-centimeter-long (17-inch) coiled and spiny penis hanging from between its lifeless flippers. Now stored for good in the museum's collection, the specimen was the first one found with its record-setting penis extruded. A year earlier, McCracken had studied seven dead *Oxyura vittata* and had already noticed that their penises were unusually long, but those, being tucked away inside their bodies, made it hard to get a good estimate of the actual length.

Celebrated bird-sex researcher Tim Birkhead of the University of

Sheffield recalls his thoughts when he first noticed this paper: "Breathtaking! My first thought was, What about the poor female? Where does all that go?"

Birkhead knew that, although extreme, the length and complexity of the Argentine lake duck penis was by no means unique among ducks. While most birds copulate by simply pressing their cloacas against one another, ducks are among the few that have a real penis, which, once the cloacas are pressed together, is everted into the vagina under lymph pressure. Along it runs a groove that carries the sperm from the cloaca deep into the female's inner sanctum. Presumably this system evolved because ducks mate in the water, where semen would wash away if they copulated with a simple cloacal kiss. In fact, at duck farms, where the birds have no option but to mate on dry land, a recurrent problem is that many a drake loses his penis because the females mistake them for a postcoital worm snack.

Among duck species, size and shape of the penis vary a lot. The harlequin duck, for example, is very modestly endowed, with a 1.5-centimeter-long (0.7-inch) penis, whereas the pintail—considered the pinnacle of duck endowment until McCracken published his Argentine lake duck photo—is blessed with a 19-centimeter-long (8-inch) coiled and spiny one; the mallard has one that is a bit shorter but similar in design. The vaginas of female ducks, however, had always been claimed to be short tubes—simple, mundane, and nothing to write home about.

At the time that Birkhead was still reeling from the photograph he saw in *Nature*, a postdoc named Patricia Brennan was working in his lab, and Birkhead suggested to her that she look at the female side of the duck story. Birkhead: "She turned out to be a master at dissecting. Halfway through her studies she called me over and she said, 'Look what I've found!'" What Brennan showed him was that the vagina of a female mallard was just as complex as the male's penis, with blind-ending side branches and twists and turns. Intrigued but skeptical, Birkhead called his friend Jean-Pierre Brillard in France, who had a lot of duck dissecting under his belt. Birkhead asked him, "Have you ever seen this?" to which Brillard replied, "*Non!*" and went off to check a few specimens of his own. "In five minutes he called me back and exclaimed, 'You're absolutely right!'"

In 2007, Brennan, Birkhead, and their collaborators published in the journal *PLOS ONE* the results of their anatomical investigations of sixteen different species of ducks, which showed that the complexity of penis and vagina went hand in hand: whenever the penis was large and curly, so was the vagina. Two things revealed that this is likely to be the outcome of male-female evolutionary competition over who has the final say in fertilization. First, the species in which rape was rife were also the ones with the most complex genitalia. And second, as the team reported in 2010 in the *Proceedings of the Royal Society B,* the corkscrew shape of the vagina makes it harder for the penis to penetrate, because whereas the vagina is always coiled clockwise, the penis coils counterclockwise. By flexing the muscles in her vagina wall, the female would be able to block a rapist male from properly inflating his countercoiled penis into her vagina.

The intrepid Brennan carried out a crucial experiment to back up these results. At a California Muskovy duck farm, she allowed drakes to mate with females, but at the cusp of the male's explosive penis eversion, she would quickly replace the female's cloaca with a glass tube. Though lined with lubricant on the inside, these tubes were otherwise designed to study how hard it was for the penis to unfold properly into the coiled vagina. And indeed, although straight or counterclockwise coiled tubes posed little problem, a glass tube resembling the female's vagina, with its clockwise coil, proved impossible to penetrate, with the penis usually folding onto itself, getting stuck halfway, or flubbing altogether. Clearly, the more elaborate penis-vagina pairs had evolved in sexually antagonistic coevolution: a mutual series of steps in which any rapist intentions of the male had been countered by vagina blockades in the female.

If the sexual ordeal that these female ducks endure sounds brutal, it is a minor inconvenience compared with the traumatic sexual experiences that beset the lives of certain invertebrates. And the term "traumatic" must be taken literally here, deriving from the Greek word for "wound." Traumatic insemination is the term reserved for those animals that, somewhere along the evolutionary line, have found a way for the male to bypass the whole female system of valves and controls altogether, by injecting sperm directly through the female's skin!

Also termed "hypodermic insemination," such drastic measures have evolved in leeches and earthworms, as well as in strepsipterans (tiny insects that parasitize bugs and bees), banana flies, several families of bugs, and the aptly named *Harpactea sadistica*. This latter species, a small spider

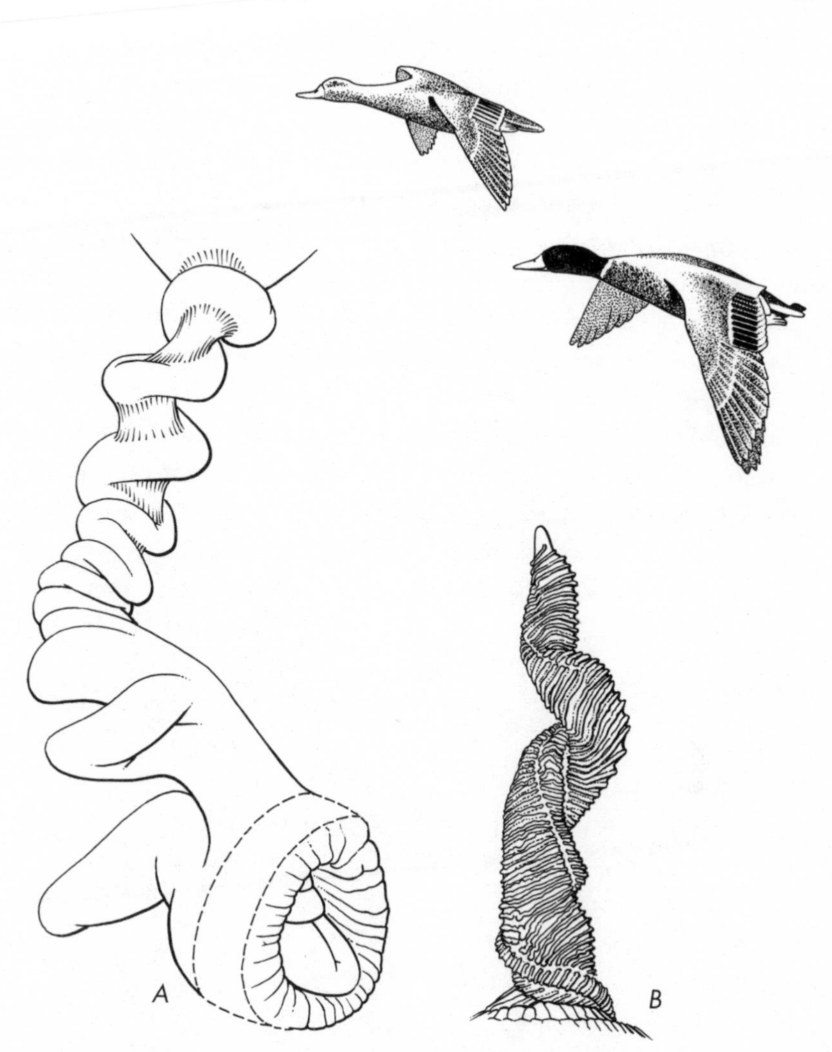

Resisting rape. The clockwise-coiled vagina of a female mallard duck (A) blocks entry by the counterclockwise penis (B), unless the female relaxes her vaginal muscles.

....

that doesn't make a web but hunts for insects on the floor of dry forests, was discovered only in 2008 in a nature reserve near Jerusalem. Most *Harpactea* species mate like all the spiders we have encountered so far in this book: the male charges its pedipalps with semen and pipes this into the genital openings of the female.

Not *Harpactea sadistica.* When a male of this species encounters a female, the pair adopts a mating position in which the male cradles the female, belly up, in his four long front legs, and then, with surgical precision, holding one pedipalp steady with the other, pierces the female's abdominal skin and injects his sperm straight into her body cavity. Not content with a single jab, he then zigzags down her tummy, creating a double row of six to eight neat punctures, changing pedipalps each time as he shifts from left to right. The female, whose sperm storage organs are rudimentary, has her eggs fertilized by these sperm, which swim straight through her body to reach the eggs in her ovaries.

Even more infamous is the traumatic insemination that is practiced by cimicids, blood-feeding bugs to which also *Cimex lectularius,* the common bedbug, belongs. Unforgettable to anyone who has ever been unlucky enough to spend several nights in bedbug-infested sleeping quarters, they will be truly memorable once you have learned about their sex lives. Living

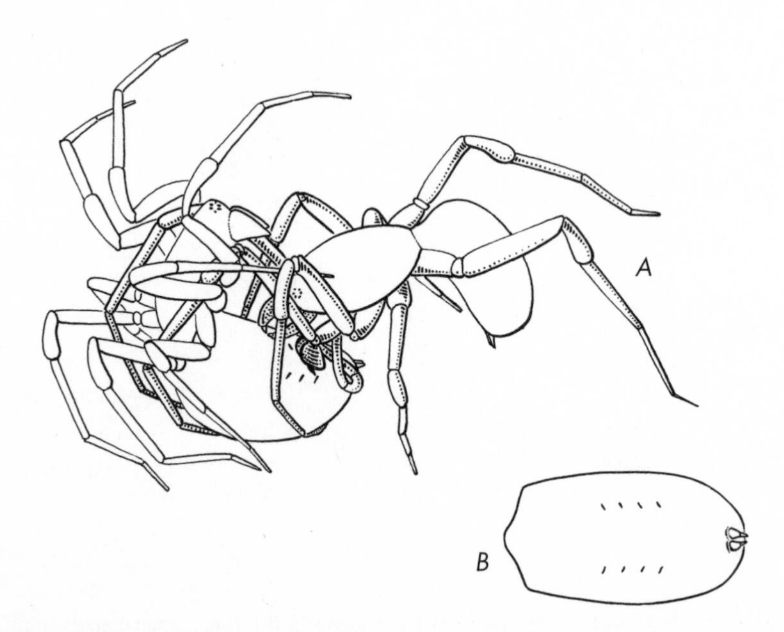

Traumatic sex. A male (A, right) of the spider *Harpactea sadistica* does not bother with copulation. He cradles the female (A, left) in his forelegs and then injects sperm directly into her belly, piercing the skin with his needle-like pedipalps, leaving a neat pattern of eight punctures (B).

in densely packed colonies in crevices near the sleeping place of their "host," sexual encounters are frequent, quick, and literally stabs in the dark. Bedbug researcher Mike Siva-Jothy of Sheffield University says: "When a female has not fed, she can avoid copulating males. But when she's fed and bloated, she's a sitting duck. There's no courtship—it's brutal in every sense of the word."

Males find their mates by a simple sit-and-wait strategy. Whenever anything resembling an engorged female lumbers past, a male bedbug will climb her right flank, push ahead until his head falls over her left shoulder, and then forcefully drive his syringe-like penis into her. (Ironically, the "penis" is actually a modified paramere, the brush-like organ that in other insects is used for gentle tapping and stroking.) Although the female has a perfectly fine vagina located at her rear end, this is of no interest to the male, as the vagina's only reproductive function is to lay eggs, not to receive any sperm. Instead, the male injects his sperm directly into the female's body by poking his penis through her skin on the right-hand side of the fifth segment of her abdomen. The sperm cells then squeeze their way through her body, even wiggling through individual cells, eventually finding the ovaries, where they fertilize ripe egg cells. There are even some unsubstantiated reports, involving other bedbug species, that males sometimes inseminate other males, where the sperm then reach the testes and proceed to be transferred, together with this male's own sperm, to the next female *he* traumatically inseminates! If true, this would be bisexuality and cuckoldry rolled into one.

In a way, such traumatic insemination is the ultimate insult to female cryptic choice. The male forgoes courtship, genital rubbing, and any negotiations of vagina, spermatheca, and other hurdles put in place by the female. Breaking all the rules of the genital game that we have come across in this book so far, he gets his way by crude copulatory treachery. And the females suffer the consequences. When a single female is housed with as many as twelve males, she is often dead within a day from the many wounds inflicted by her suitors' penises. And even a single male can shorten a female's life span considerably. Alastair Stutt, a student of Siva-Jothy's, did experiments in which he let females mate either with regular males or with males in which he had rendered the penis harmless by sticking it to its belly with a droplet of superglue. He found that the females housed with these incapacitated males lived for more than four months, whereas the females that had to suffer functional males survived for only eleven weeks or so.

But all is not as it seems. Female bedbugs have evolved a complex system to cope with the worst effects of traumatic insemination. First of all, they carry a so-called spermalege. This amounts to a brand-new set of genitalia, developed in the site where the male normally impales her. In *Cimex lectularius,* this includes a slit-like organ, effectively a second vagina, on the right-hand side of her belly, and when the male adopts the copulatory position and launches his stylet-like penis into her flesh, his penis is usually guided into it. Underneath this pseudo-vagina lies a large organ, the mesospermalege, which has two functions. It receives the male's ejaculate, guides the sperm cells into the blood, and sends them on their way toward her single ovary. It also has a sanitary role: the grimy conditions under which bedbugs live mean that a male's pointed member is covered in harmful bacteria and fungi that it inoculates the female with as he inserts his penis into her. To deal with this, the mesospermalege houses a massive immune system that neutralizes most of these sexually transmitted diseases. Entomologists have done experiments by stabbing female bedbugs all over their bodies with infected replica bedbug penises. (Siva-Jothy: "There's actually a guy in my lab whose pleasure it is to produce glass penises.") They have discovered that when the females are jabbed into their spermalege, their lifetime egg output is not affected, but when they are stung in other parts of their body, they get so diseased that their offspring production is much less.

In a different kind of bedbug, *Stricticimex* ("batbug" would be a better term for them, since they feed on the blood of bats in caves), the mesospermalege has evolved into such a complex system that traumatic insemination effectively has become regular insemination again: rather than traveling like unguided missiles through the female body, the sperm, after having been squirted into the tissue underneath the pseudo-vagina, are directed via a set of tubes and storage chambers to the ovaries, and the female in fact regains full control of her sex life again. So, just like the long and coiled penises and vaginas of ducks, the evolution of traumatic insemination in bedbugs also seems to be the product of a series of male adaptations to bypass cryptic female choice along with female countermeasures to gain the upper hand again.

Although traumatic insemination sounds extremely bizarre and eccentric, the fact that it has appeared in so many unrelated kinds of animals means that it is not that unimaginable a route for evolution to take. This probably has to do with the fact that sperm are already designed to wander foreign territory and search and penetrate (egg) cells. It is a rather small step from doing that within the legitimate confines of the female reproductive tract to doing it cross-country, as it were. In fact, free-swimming sperm are also found in the female bodies of animals that do *not* normally perform traumatic insemination, having apparently escaped from vagina or spermatheca. In certain mites, some sperm escape from the female genitals and are encountered traveling aimlessly in the female mite's blood. And already in the 1950s veterinary researchers discovered the very effective way of artificially inseminating female guinea pigs and chickens by simply putting sperm in a hypodermic needle and injecting it directly into the animal's belly!

The evolutionary distance from fellow vertebrates to ourselves is small. Indeed, human sperm are also sometimes caught escaping from the reproductive system and wandering the body cavity of human females. To estimate how often this happens, a group of three American gynecologists made clever use of a rare congenital condition known as "noncommunicating uterine horns." In the embryo, the uterus develops from two tubes, known as "Müllerian ducts." In boy embryos these gradually degenerate, while in girl embryos they grow and fuse into a single uterus with two fallopian tubes. In about one in four thousand women, however, the Müllerian ducts fail to fuse properly, leading to a uterus that is split down the middle into two chambers—or "horns"—only one of which connects to the cervix and the vagina.

Gerard Nahum of Duke University Medical Center and two of his colleagues surveyed the medical literature for pregnancies in women with such noncommunicating uterine horns. They found that a fetus was just as likely to be found in the chamber of the uterus that had a connection with the vagina as in the one that had no such connection. The inescapable conclusion was that all these (obstetrically problematic) pregnancies in the "blind" uterus chamber must have been caused by sperm cells migrating through the other chamber, up the fallopian tube, then into the woman's abdominal cavity, entering the other fallopian tube from the back entrance. There, one of them fertilized a ripe egg cell, which then embedded itself in the blind uterus chamber.

The staggering implication of this study is that in all women—including those without this congenital condition—sperm might be traveling outside of the accepted reproductive avenues all the time, taking the shortcut via a woman's belly if there are eggs to be fertilized in the other fallopian tube over yonder. When you think of it that way, the traumatic insemination in bedbugs with their intrepid innards-traversing sperm cells is perhaps not that foreign after all.

Traumatic insemination evolved because it helps ejaculates to elbow their way in front of other males' sperm depositions. Just like the other tricks we have witnessed in this chapter—forced copulation, sperm flushing, scooping, and scraping—it is a way in which evolution has dealt effectively with the clear and present danger of other males simultaneously vying for a fertile female's favors. Clearly, we are well within the territory of sexually antagonistic coevolution now. But there is more. Current sperm competition is not all that a male has to worry about. Since females usually store sperm for a long time, future suitors are just as much of a headache. In the next chapter, we will meet yet another category of sexually antagonistic tricks up evolution's sleeve—genital innovations that hamper such subsequent suitors' success.

Chapter 7

Future Suitors

Tucked away in a far corner of the northern German lowlands on the shore of the Baltic Sea, the small university town of Greifswald is a harmless vestige of Hanseatic commerce and rows of preunification apartment buildings. Its main claim to fame lies in being the birthplace of the Romantic-era painter Caspar David Friedrich. And as my old VW Beetle trundles its way along Greifswald's access roads, snow still piled high on the sidewalks, nothing in the townscape (and certainly not the many austere Reformed Church spires), immortalized two centuries ago by Friedrich, betrays that the city is the hub of German genitalia research.

Up in the university's bird observatory I locate, rather paradoxically, the office of snail-genitalia researcher Martin Haase, whose name will come up in the next chapter, on hermaphrodite sex. And downtown, the two buildings of the Greifswald Zoological Institute house a group of experts on arthropod sex. My first appointment is with retired professor Gerd Alberti and his colleague Antonella di Palma, who receive me in the institute's lecture theater. Accidentally comical, kindly, and speaking in a husky, high-pitched voice, Alberti hands me a CD that, as he casually remarks, contains his life's work: 242 scientific papers on the reproduction of mites, spiders, and a few other kinds of invertebrates. He sits me down on one of the theater's benches and then unleashes a one-hour travelogue through the amazing world of mite and spider sex, starting each new episode with the exclamation "And this is really very funny and exciting!"

One of those funny and exciting points: spiders and mites are rather unusual in that the females often have separate body orifices for sperm input and egg output. Whereas most animals have to use a single multipurpose orifice for both procreational activities, many mites (especially the ones Alberti and Di Palma have been studying) receive sperm through small openings at the "hips" of their legs and yet lay their eggs through a different opening at their rear end. And most spiders practice the same division of labor: the males empty their pedipalps into a female spider's two insemination ducts, which flank a third, egg-laying opening in the middle of her abdomen. The significance of this dawns on me only as I cross town toward the institute's other branch, where institute director Gabriele Uhl has her office.

For many years, Uhl, a charming and cheerful arachnologist, has been studying so-called mating plugs in spiders. Much to the annoyance of spider taxonomists, who need a clear view of the female genitalia for proper identification of the species, spider females' private parts are often clogged with "dirt" or unidentifiable bits and pieces. On closer inspection, this debris often turns out to consist of male body parts—more precisely, tips of their pedipalps. In a very large proportion of spiders, Uhl says, after copulation a male breaks off a part of his own genitals and leaves it behind in his mate's. Not as an unsavory souvenir or as the result of an accident during his hasty getaway, but rather in an attempt to secure his paternity.

Spider researchers suspect that these bits of pedipalp obstruct the female's genitalia against penetration attempts by subsequent suitors. And, says Uhl, the fact that spider females usually have separate insemination and egg-laying openings is one reason why spiders are "preadapted" for the usage of such arachnid chastity belts; after all, unlike in many other animals, the plug will not be in the way once the female starts giving birth to spiderlings. Sure enough, plugging happens everywhere in the animal world, but it proves particularly popular among spiders.

But Uhl raises her index finger and allows her eyes to twinkle. "Ah, when we say 'plug,' this does not mean that they *are* plugs," she warns. In the black widow spider, for instance, a pedipalp tip in the female genitals does not fully prevent later males from mating with that female as well, since up to five pedipalp tips can be found lodged in a female's genitalia. This means that the first mate's "plug" does not dissuade numerous subsequent males from mating with the female and leaving pedipalp tips of their own. A male with broken pedipalps is effectively neutered—they never grow back—so why would males perform such acts of self-mutilation if they do not completely prevent copulation by other males? To find out the answer to this question, Uhl and her student Stefan Nessler and colleague Jutta Schneider have been studying mating plug effectiveness in the wasp spider.

The wasp spider, or *Argiope bruennichi,* is a striking orb-web-building spider, easily recognized by its main design feature of black and yellow cross stripes on its body as well as its habit of hanging in its webs with its eight legs held two by two in an X shape. In recent years, probably thanks to climate change, the large spider has been staging an invasion of the UK, the Netherlands, Denmark, and northern Germany from its original stronghold in Central Europe, bringing the spectacular species more urgently to the attention of researchers like Uhl.

She gets up and walks across her office to look for the research project papers. Scanning the shelves, flipping through heaps of papers, and sighing melodiously ("Hhhhhmmm, it's gone. Oh, no—here it is!"), she returns with a small pile of student reports and reprints of published papers, which she fans out in front of me. Pointing at a color photo showing a closeup view of a male wasp spider's pincer-like pedipalps, she explains that they used CT scanning to work out precisely how the male uses his pedipalps to grasp the female genitalia, then reaches deep inside her and pumps his sperm into her, and finally snaps off the tip of one of the pincers' prongs at a predefined breaking point, leaving behind a sizable chunk of his anatomy in the female after copulation.

To work out what benefit the male gains from crowning his copulation with an act of self-neutering, Uhl and her team first visited meadows and picked almost three hundred immature males and females from their webs. Letting each spider grow up in its own plastic cup in the lab, they generated a large number of mature virgin males and females. Then, in Plexiglas cages, they gave some their first sexual experience by releasing a male into the web of a female. The male wasp spider is much smaller than the female. He mates with her by crawling between her and her web and then, belly to belly, going through the ritualized motions of emptying either his right pedipalp into her right spermatheca or his left one into her left spermatheca, the whole process usually lasting only seconds. Under natural conditions, coitus then ends rather abruptly with the female sinking her fangs into the male, swiftly wrapping him in silk, and eating him. But in the lab, the researchers rescued the males by snatching them from the females' claws as soon as his pedipalp tip had broken off and copulation was done.

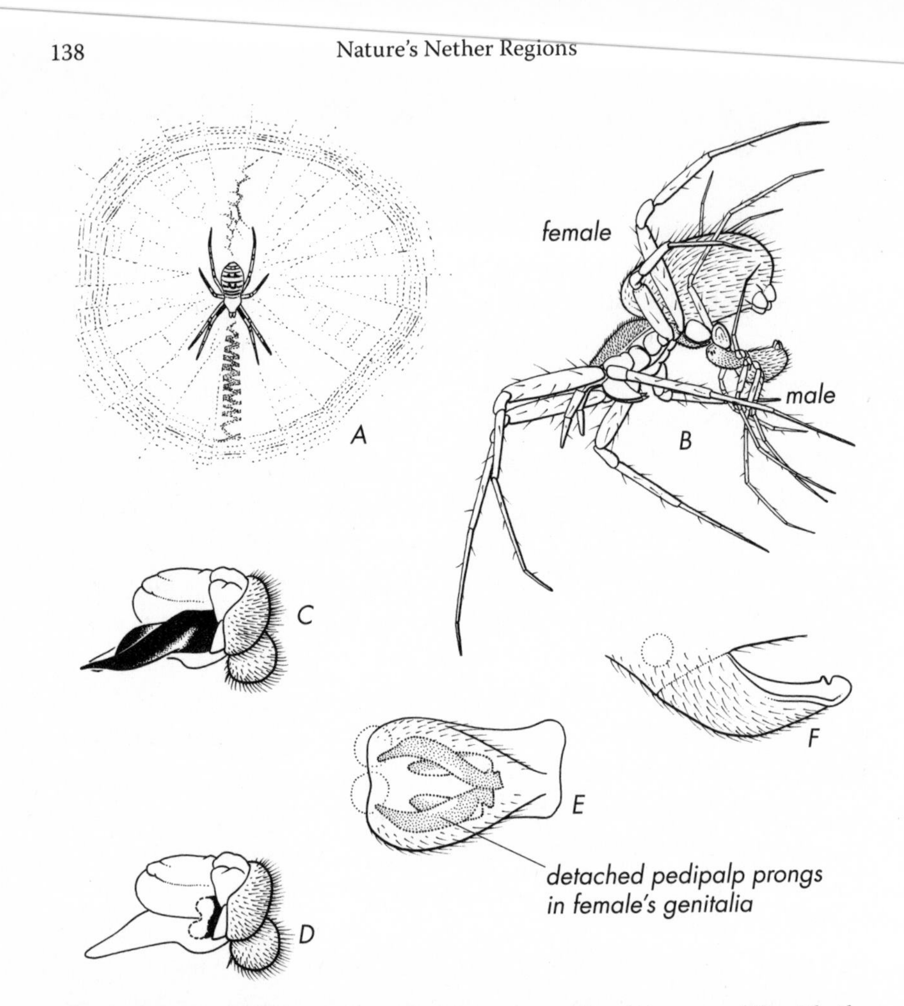

Chastity belts. When the large female wasp spider (A) mates (B) with the small male, a prong on his pedipalp breaks off and is left behind in her genitalia. C shows a complete pedipalp; in D the prong is broken off; and E shows two detached prongs clogging a female's genitalia, with a side view shown in F.

....

Having thus created a collection of females, some with, others without one of their insemination ducts plugged, they then turned to another batch of fresh males and proceeded to amputate one of their pedipalps to create a battalion of males with either their left or their right pedipalp missing. Now the stage was set for a clever experiment. The spiders' mating behavior is so predictable that a male will *always* insert his right pedipalp in a female's right insemination duct, and never into her left one, so the researchers could engineer sexual encounters in such a way that they could be sure that the duct a male was inserting his one remaining pedipalp into was either plugged by a previous male or still vacant. Then they measured how long the male kept his pedipalp inside the female. And since sperm is injected into the female with an ongoing pumping action, the longer the pedipalp remained inside, the more sperm was transferred.

Sure enough, the results showed that plugging the female's genitalia, though not completely preventing a second male from mating with her, did make it harder for subsequent males to do their thing. A male forced to put his pedipalp into an already plugged insemination duct left it in there only half as long as a male ejaculating into a vacant hole. This means that to a wasp spider, there is a benefit to leaving part of his pedipalp behind: it does not stop later males from also inserting their pedipalps, but it does hamper them in ejaculating with full force. A male that successfully plugs a female's genitalia therefore has a higher chance of getting his just deserts than a male that doesn't plug. And because many males do not survive their defloration anyway, they gain much but lose little by such self-emasculation.

So, the generally poor prognosis for any male spider's sex-life expectancy conspires with strong competition among males to produce another reason why spiders are preadapted for evolving mating plugs. Faced with such a high chance that your first time is also going to be your last, converting your genitals into a paternity-securing device is a small sacrifice given the benefits gained from protecting yourself from posthumous sperm competition.

Chapter 5's *Tidarren* spiders, which twist off one of their own massive pedipalps the better to move about, are another example of a spider that has opted for plugging and self-sacrifice. Since his single remaining palp can be used only once, and since he will die a certain death in the female's embrace, *Tidarren argo* uses his *entire* pedipalp as a mating plug. This works as follows: as soon as he has begun pumping his sperm into the female, she then grabs him and twists him round and round until he gets detached from his securely fastened pedipalp, which continues to actively pump sperm into her for several hours after having been detached. As the female snacks on his body, he rests assured in the knowledge that his pedipalp will keep his sperm in and prevent other males from adding theirs. Barbara Knoflach of the University of Innsbruck, Austria, who has been responsible for discovering many of these details about *Tidarren*'s bizarre sex life, ran experiments to test the effectiveness of such a plug made from a whole pedipalp. She introduced fresh males to females that had a previous male's pedipalp still stuck to their epigyne and discovered that these females tended to be barred from copulating again as long as the plug was in place.

While we're at it, *Tidarren* has more surprises in store in the mating plug department. Knoflach discovered an even more extreme case in another species, *Tidarren sisyphoides*—the same species for which Margarita Ramos proved that having one pedipalp is less cumbersome than having two. In this species, it turns out that the male uses his *entire body* as a chastity belt! As soon as copulation has begun, the puny male, his oversized pedipalp happily pumping away, stiffens and dies—just like that. The female does not cannibalize him, but for hours after mating walks around with her dead lover affixed to her genitalia. Only hours later does she remove his corpse-cum-plug and discard it.

Since in most animals males normally hope to use their genitals more than once, not many animals besides spiders have evolved mating plugs that require the amputation of parts of their penis or even of their entire penis. But there are other ways to produce a mating plug. One option: add two-component glue to your ejaculate.

Solid Semen

A well-known zoologist once said, "If you want to discover something new, then don't read old German literature." And indeed, the time-honored Central European tradition of meticulous dissection and painstaking monographic description of anything that walks, crawls, swims, or slithers also produced the very first discovery of a glue-like mating plug. Back in 1847, Leipzig zoologist Karl Georg Friedrich Rudolf Leuckart first winkled what he called a "vaginal plug" from the pudenda of a guinea pig. As we now know, in this and many other rodents, immediately after ejaculation part of the male's semen solidifies inside the female and turns into a solid plug that forms a perfect cast of her vagina and remains stuck there for up to several days, rendering her genitalia inaccessible for the rest of the fertile period of an estrus cycle.

And not just in rodents. Insects, shrimp, nematode worms, snakes and lizards, and many other mammals besides rodents produce such plugs. This includes primates. In 1930, a scientist named Otto L. Tinklepaugh was the first to discover that even our closest relative, the chimpanzee, makes solid sperm plugs. Working in Yale's primate center, he was studying the sexual cycle of one female chimp and her consort by flushing her vagina daily with a blunt-tipped syringe and doing smear tests on the effluent. During the female's estrus, often his syringe would get clogged by strange bits of hard material. In the end, he managed to retrieve from her vagina a hard, solid plug measuring 7 centimeters long and 1.5 centimeters thick (or 2.75 by 0.6 inches). "Shaped much like the finger of a glove," Tinklepaugh wrote in the short article in the *Anatomical Record* that he devoted to the matter.

Since Tinklepaugh's days, primatologists have taken to observing up close the sex lives of all imaginable monkeys and apes, and we have already come across a few memorable examples in this book. The spoils of these decades of sexual scrutiny include a veritable catalog of mating plugs in lots of primate species, and in 2002 Alan Dixson and Matthew Anderson of the San Diego Zoo published a compilation. Even among our closest relatives, the great apes, three different types of plugs could be discerned. First, there were the solid finger-like plugs produced by chimp and bonobo semen. Second, in orangutans, semen in the female's vagina coagulates into rubbery globules but does not form a single solid plug. And third, in gorillas and humans, semen congeals into a squishy, shapeless gelatinous mass. Mapping these three types across all forty primate species in their list, Dixson and Anderson saw an interesting pattern: the more promiscuous the females were known to be, the more likely it was that the male left behind a heavy-duty mating plug of the chimp/bonobo type—a sure sign that plugging, also in mammals, is a way for males to counter sperm competition.

But how does liquid semen manage to turn into something with the consistency of putty? Fortunately, decades of research into male (in)fertility have given us a very detailed insight into the various proteins, salts, sugars, and other constituents of semen and how they work together. Allow me a brief recap of mammal male sexual function.

The sperm cells are, of course, produced in the testicles and, when half-baked, are moved to the epididymis, a tightly coiled tube (stretched out, it would be six meters long in humans!) that is packed into a slug-like lump at the back of each testicle. While moving along the length of these six meters, the sperm cells mature and accumulate, poised for action, in the tail part of the epididymis. On the verge of orgasm, tight packs of these mature sperm cells are squeezed out of storage and pushed along the thin tube (vas deferens) that passes around the bladder and then, behind it, meets the seminal vesicles. These glands sit at the confluence of the two vas deferens tubes coming from both testicles, and they produce some two-thirds of the liquid of the semen. They also add a cocktail of proteins, including so-called semenogelin. Remember that name.

Having received the seminal vesicles' contribution, and orgasm now well under way, the semen rushes on via the vas deferens tubes (which here and beyond are fused into a single tube) to meet the next gland, the prostate. This adds some additional liquid of its own, containing, among many other constituents, an enzyme called transglutaminase 4, or *TGM4.* But rather than mixing the prostate-derived component with the rest of the semen, tight coordination of the muscles in the walls of the prostate and the vas deferens, which by now is called the ejaculatory duct, ensures that the portion added by the prostate remains a separate wave riding in front of the droplets released by the seminal vesicles. Finally, the stream of semen passes along a pair of small glands called Cowper's glands, which have been busy oozing clear lubricant into the ejaculatory duct since well in advance of orgasm, and much of which is added to the bow wave of the semen as it picks up speed at the base of the penis and is finally ejaculated in a series of pulses caused by those familiar 0.8-second genital spasms that we met in Chapter 4.

Biochemists have made use of the spasmodic nature of ejaculation by catching separate ejaculate droplets in separate cups and studying the constituents of such "split ejaculates." It turns out that, thanks to the successive contributions of the various glands, the first few drops of ejaculate contain mostly sperm cells and semen produced by the prostate and the Cowper's glands—including *TGM4*—whereas the later droplets are poor in sperm cells but rich in seminal vesicle fluids and therefore contain a lot of semenogelin. Once inside the vagina, *TGM4* and semenogelin come in contact with each other and start a biochemical interaction. Like all proteins, semenogelin consists of a long chain of amino acids. *TGM4* specifically singles out one kind of amino acid (called glutamine) in these chains, and connects this to another amino acid, lysine, in neighboring semenogelin molecules, creating a tangled web of cross-linked chains. It is the same process that is responsible for the toughness of our skin. And it is what causes the solid mating plug to form.

Obviously, semenogelin and *TGM4* should not meet each other before it is time for a mating plug to form, which explains why they are produced and stored by two separate glands inside the male's body. Like the two constituents of two-component glue, they mix only when squirted out. Now, genes to produce semenogelin and *TGM4* are present in the DNA of all apes and humans, so why the big difference in plug consistency? The answer lies in the exact structure of those DNA blueprints.

While still an undergraduate student at Brown University, primatologist Sarah Kingan did a research project on the DNA code for semenogelin in humans, chimpanzees, and gorillas. She discovered that the code in gorillas was so garbled that it had stopped producing a functioning protein. And Sarah Carnahan of Duquesne University in Pittsburgh found that the same was true for *TGM4*. This makes sense, since gorilla males monopolize their females, so their semen rarely, if ever, needs to compete with that of other males—mating plugs would be superfluous. No wonder that mutations rendering their glue components useless could have accumulated in these apes without doing any evolutionary harm. In fact, Carnahan found this held true in the siamang gibbon, a smaller species of ape that fosters long-term monogamous pair bonds, also freeing their semen proteins from the task of dealing with rival sperm.

As Kingan and Carnahan discovered, however, the situation is different in chimps, bonobos, orangutans, and humans, where sperm competition is a greater risk (although we like to think of ourselves as basically monogamous, humans are still much less faithful than siamangs). In these species, the genes are all still functional. Not only that, but especially in chimpanzees and bonobos, the DNA sequences show telltale signs of rapid evolution. Their semenogelins have become much bigger, creating additional lysine-glutamate links to be forged by *TGM4* and, hence, producing a more sturdy mating plug. The story does not end there, though, because recently the female side of the story has also become clearer—albeit in mice, not men.

In a 2013 article in the journal *PLOS Genetics,* Matthew Dean of the University of Southern California gave a new spin to the mammal mating plug story by using mouse knockout. Rather than an unfair kind of boxing match, "mouse knockout" refers to the geneticist's trick of disabling certain genes in lab mice. There's even an International Knockout Mouse Consortium (don't laugh), a worldwide collaboration to create families of genetically engineered mice that lack certain genes. For many of those genes, even though the geneticists know exactly how to knock them out, the precise function is still unknown. Finding out those functions is what knockout mice are for, by studying which faculties are impaired in a mouse with a particular gene knocked out. What Dean did was ask the IKMC for a couple of mice that had had their *TGM4* knocked out and set up tests in the lab to compare these males' mating success with that of regular mice.

As expected, the semen of mice lacking *TGM4* proteins failed to congeal to a proper mating plug inside the vagina of Mrs. Mouse. But in addition to making the male prone to sperm competition from other males, the absence of the *TGM4*-induced plug seemed to have more effects. Dean found that far fewer sperm managed to migrate up the female's reproductive tract after insemination with *TGM4*-less semen than with regular, coagulating semen. This means that, in addition to blocking subsequent males, the mating plug may also help prevent sperm dumping by the female. Dean discovered that of the *TGM4*-lacking sperm that did make it to the fallopian tubes, fewer pregnancies resulted than would normally be the case. So apparently a mating plug also prevents active embryo abortion by the female (see Chapter 4), which, says Dean, may be due to continued "physical stimulation" of the female by the plug long after the male's penis has disappeared from her vagina.

Again, we see that a mating plug, which superficially seems just another nasty trick by the male to enforce chastity in his female, may actually have a more subtle role. Yes, it makes it harder for subsequent suitors to add their sperm, but at the same time a plug may prevent sperm dumping *and* be a postcoital echo of the penis's sensory courtship. Moreover, as we saw in the *Tidarren* spiders in which the female helps the male detach his pedipalp to act as a chastity belt on her epigyne, females sometimes cooperate in letting a male plug her genitals.

Research in squirrels has likewise shown that the female's role in the whole plugging business may not be passive. In two consecutive winters, zoologist John Koprowski woke every morning before dawn to observe the morning mating rituals in the two species of tree squirrels on his University of Kansas campus grounds. During the winter mating season, gangs of squirrel males hang around impatiently near the entrance of a female's den and try to mate with her as soon as she, sleepy eyed, makes her appearance. Registering copulations through his binoculars, Koprowski saw that in more than half, as soon as mating was over, the female would nibble away at the mating plug—roughly the same size and shape as a cigarette butt—or chuck it to the ground. But in all other cases a female seemed happy to keep the plug sitting in her vagina until it naturally dissolved a day or so later. Perhaps a nice and solid plug is a good thing to keep in your vagina for a little while. Either for "physical stimulation" or because a male that can lay down a decent mating plug is the kind of father you'd like for your sons . . .

Substance Abuse

Let's wallow in semen a little while longer, shall we? We have already seen that, even in humans, there is more to this substance than meets the eye. It contains proteins that, when mixed together, can forge a mating plug. It also contains sugars as sperm fuel, proteins that protect the sperm cells from the acidic vaginal environment, zinc that keeps the sperm's DNA in good shape, and chemical compounds that prevent the sperm cells from becoming overenthusiastic prematurely.

But this list of ingredients is just the tip of the iceberg. Human ejaculates are home to hundreds of different proteins (which in certain women cause a kind of "sperm hay fever," an allergic reaction to semen). And those are not trace amounts either; most of them occur in considerable concentrations, so they must be doing something important—we just don't know what. Even in the ejaculate of the lowly banana fly *Drosophila melanogaster,* researchers have identified no fewer than 133 different kinds of proteins. One hundred and thirty-three! And this excludes the many proteins that are in the sperm cells themselves. These 133 are all produced by the banana fly version of the prostate, which releases them into the liquid portion of the semen.

Fortunately, banana flies being the lab biologist's workhorse, we know quite a bit about what their seminal proteins do—or at least more than we know about their human counterparts. To begin with, we know that they evolve as fast as the genitalia that deliver and receive them. Rama Singh of McMaster University in Hamilton, Canada, has been studying the protein cocktails in different species of banana flies and has discovered that the DNA codes for these genes, while remaining functional, are constantly and rapidly changing along the branches of the banana fly evolutionary tree. "The fastest evolving proteins we know," he says. And a sure sign they are perpetually pushed around by sexual selection.

Banana fly researchers are quite confident that some of the ingredients of these biochemical cocktails are involved in a kind of neuropsychological manipulation. They hijack a female's hormonal system by shutting down her sex drive, causing her to go completely off males for up to several days after having received a load of semen. Females that have recently been inseminated start kicking away their suitors or, when harassed, extend their egg-laying tube, which blocks access to the vagina. They even begin exuding a scent that renders them unattractive. All this is induced by semen components that end up in her bloodstream. Usually, females prefer to have multiple fathers for a clutch of eggs—to generate healthy genetic diversity. But by usurping a female's decision to add other males' sperm to the mix that she is using to fertilize her eggs with, a male thus can prevent sperm competition. The whole process is akin to leaving a mental mating plug.

One of the substances that have such an "antiaphrodisiac" effect is sex peptide, a small protein molecule—small enough to pass straight through the wall of the vagina into the female's bloodstream—that is produced in the glands that sit next to a male fly's genitalia. It is present in the semen but also sticks to the tails of the sperm cells. The sex peptide that is floating free in the semen does its work quickly: even before mating is fully over, it has already seeped through the vagina wall into the female's blood, and within minutes begins to stick to receptors near her brain, where it makes the female's interest in other males plummet. For up to seven hours, the initial shot of sex peptide causes females to give males the cold shoulder. Meanwhile, the sex peptide that sits on the sperm tails is beginning to break free, sustaining a steady IV drip of antiaphrodisiac that lasts for about a week—enough time to give the sperm a free passage, unchallenged by other males' sperm.

Sex peptide is only one of the multitude of chemical compounds in semen that a female banana fly receives each time she mates with a male. What do the rest do? Well, research in other insects can give us an inkling of what such substances might be capable of. The semen of the American fire beetle *Neopyrochroa flabellata,* for example, is spiked with the poisonous compound cantharidin. Although in the human world this substance is known as the infamous aphrodisiac "Spanish fly," the benefit gained by the male fire beetle is not his mate's increased ardor, but rather the fact that the stuff ends up in the eggs fertilized by his sperm, which protects them against being eaten by ladybird beetles. Another substance, the protein PSP1, is ejaculated by the male corn earworm moth into his mate and there immediately shuts down the production of pheromones, meaning that other males (which totally rely on scent) can no longer find her. And then there's *Argas persicus.* In this tick, believe it or not, the male produces a soda-bottle-like spermatophore from his genitalia, takes it into his jaws, bites off the cap, and then sticks it, neck first, into the female's vagina. The appearance of carbon dioxide bubbles inside the spermatophore then forces out the sperm and other contents into the female. And at least one of these contents is a compound that cranks up the female's egg production rate—which may mean more offspring to be sired by these sperm.

This list makes one wonder whether some of the many proteins in human semen could also have such manipulative effects. If they do, then this would be one way to explain the results of a study by Gordon Gallup and Rebecca Burch, whom we already caught red-handed in Chapter 6 while applying a dildo to a latex vagina filled with artificial semen. In this other study, they had almost three hundred female students fill out questionnaires relating to sex and mental health. The results showed that women who always use a condom—and so are protected against the effects of proteins in the semen—score almost 50 percent higher on a scale of depression-related symptoms than women who never use condoms, which might indicate (but doesn't prove) that substances in semen interfere with the female nervous system. And there is also evidence that pregnant women who have unprotected sex with their partner during pregnancy are less likely to suffer from so-called preeclampsia than women who use a condom. Since preeclampsia is a kind of inflammation of the woman's body induced by her fetus—basically the mother becomes allergic to her own child—this might mean that substances in the semen take over part of the regulation of the woman's immune system.

It seems that semen may be full of manipulative substances: the male using the female's reproductive system to wage a chemical warfare against her reproductive autonomy or even, in an act of evolutionary paranoia, against potential future competitors. But this view of semen as a magic cocktail may be too one-sided. After all, for these semen compounds to do their job properly, they need to pass through the wall of the female's genitalia into her bloodstream. And although very tiny molecules may do so unassisted, larger proteins are too bulky to slip through the cracks in the female's tissues—they need help from the female to do so. Take the grasshopper *Gomphocerus rufus.* When researchers injected extract from one of the male's genital glands directly into a female's blood, nothing happened. But when they injected it into her spermatheca, she promptly started kicking away at any male that approached her. They traced this to compounds from the semen latching on to receptors on bristles in the spermatheca wall, which then set in motion a physiological reaction leading to this irritable behavior. So this means that the female must be mediating the action of these so-called manipulative substances. The upshot is that rather than chemical warfare, many of these semen compounds are engaged in what might better be termed chemical communication. The female's cellular apparatus needs to open molecular valves to allow the males' proteins to activate her physiology. So, again, female discretion is key.

But even at this molecular level, there are some males that won't take a chemical no for an answer. Take the housefly. This all too familiar insect is so common that you've probably witnessed, willingly or not, two of them mating on a wall—the female absentmindedly preening her forelegs. Next time, imagine the following. Within minutes after mating commences, the male places his ejaculate in the female's vagina, the liquid portion of which is received in small vaginal pouches. Ten minutes into the copulation, aggressive chemical compounds in the semen begin leaching holes in the vagina wall. The thin layer of cells begins to disintegrate and the semen starts leaking into the female's blood. This goes on for at least another half hour, by which time large, gaping holes have appeared in the vagina wall, the semen has been absorbed by the female's bloodstream, and antiaphrodisiac proteins are beginning to do their work. The male then flies off, leaving the female frigid for at least three weeks (if she can dodge rolled-up newspapers for that long).

Corrosive compounds, mind-altering molecules, subversive substances . . . in all kinds of animals, the male's ejaculate appears to contain not-so-innocent ingredients. What's the impact of such chemical onslaught on a female animal's body? Could all this male-administered medication become a health hazard for a partner? Indeed, in some animals the consequences of semen on a female's health may be dire. Back in 1995, British evolutionary geneticists Tracey Chapman and Linda Partridge discovered that repeated exposure to male semen reduces a female banana fly's life expectancy by 20 percent. They proved this by comparing the impact of regular male flies with that of males genetically engineered so that they could produce sperm, but none of the seminal proteins. Females that mated with the latter kind of males survived for more than forty days, whereas the ones whose bodies had to endure the male's chemical warfare usually keeled over by week four.

Now, toxicity of substances sloshing around in a male's semen sounds paradoxical. Why would a male jeopardize the health of the female that is to bear his children? Researchers Alberto Civetta and Andrew Clark provided the answer. They discovered that some male banana flies are genetically predisposed to having more harmful semen than others. And those males were also the ones more successful at fathering a female's babies. In other words, the more noxious semen was also better at persuading the female to use the sperm cells in it for fertilizing her eggs. So despite being his mate's ultimate downfall, a male with particularly nasty ejaculate would still reap reproductive benefits. On the female side, however, there is also something going on. In seed beetles, which likewise fill their ejaculates with female-unfriendly substances, some females have more genetic immunity to the toxic effects than others.

With male genes that make semen proteins differ in their toxicity, and female genes that cause variation in how susceptible a female will be, the stage is set for rapidly spiraling evolution, in which both sexes will continuously evolve to deal with semen in the way that best serves their own interests. And, as Bateman has taught us, those interests are not one and the same. The outcome, on a longish timescale, is a sort of evolutionary tango, with any male step (add manipulative protein X to the mix) answered by female countersteps (evolve an antidote to X). This explains why Rama Singh found that those semen proteins are in the fast lane for evolution, because there is a constant pressure to change and adapt the blueprint for these proteins—something that is true not just for flies, but for most animals.

But what if males do not only evolve more efficient proteins to spike their semen with, but also more efficient ways of administering it? That is when spiking takes on a painfully literal meaning.

Love Hurts

"Well, that's about it," says Gabriele Uhl as she slaps shut the latest student report she had pulled off her shelf and adds it to the sprawling heap of books, reprints, and spiders-in-spirit that has grown on her desk over the course of our interview. We have meandered from wasp spiders' mating plugs to what goes on in the mind of a copulating female cellar spider and along the way touched on the constituents of spider and other semen. And just as gaps begin to fall in our conversation, there is a knock on the door and in breezes Uhl's husband, beetle enthusiast Michael Schmitt, who briefly pumps my hand and then drops himself into a desk chair in the corner. "He just had to give an exam," Uhl explains to me, with a wink. "Which he only remembered minutes beforehand."

After reviving himself with a cup of tea, Schmitt gets up, grabs his coat, and takes me downstairs to his own room to chat about spiny beetle penises—as promised.

One of the things Schmitt shows me in his cozy, bookshelf-clad office is an article one of his students, Lasse Hubweber, wrote on longhorn beetle penises. As we have seen before (when discussing René Jeannel and his cave beetles), the beetle penis usually consists of a tough capsule from which it can conjure an inflatable internal sac, and longhorn beetles are no exception. Like cave beetles, the internal sac of the longhorn beetle penis

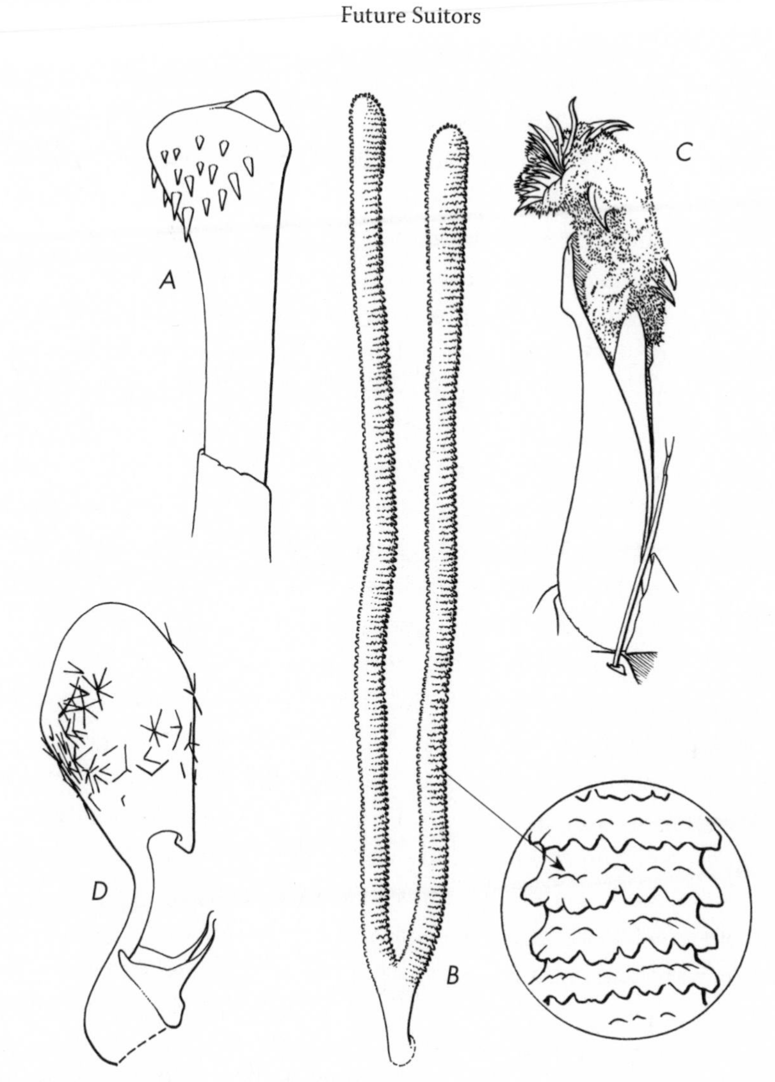

Ouch! Spiny penises occur throughout the animal kingdom, including in bush babies (A) and king cobras (B) as well as beetles (C) and moths (D), with their caltrop cornuti.

....

is studded with painful-looking spines and bristles. The electron microscope photos in Hubweber's article don't leave much to the imagination. The penis of the graceful European species *Alosterna tabacicolor* carries rows and rows of shark-tooth-like backward-pointing spines. And the black-and-white Southeast Asian *Chlorophorus sumatrensis* can extrude a ghastly-looking, rasp-like tube with solid outward-pointing teeth. "Quite something, huh?" says Schmitt.

Of course, having spines on your penis is nothing peculiar to longhorn and cave beetles. Many beetles have this, as do banana, carrion, and caddis flies, and certain larger animals, too—in this book we have already come across the spiny penises of elephant shrews, rodents, and ducks. Then there are the colubroid snakes, proud owners of two penises, each a fascinatingly beautiful bouquet of red and deep purple grooved, multitoothed spikes, like a clutch of medieval halberds. Hoary bats sport long needles on their glans that are longer than the glans itself. And finally the mysterious but aptly named "caltrop cornuti" on the penises of certain moths. These star-like multipronged thorns, similar in shape to the caltrops scattered on the ground in warfare to slow down pursuing armies, are detached from the penis during mating and stay behind in the female moth's genitals.

Nobody knows what sinister purpose caltrop cornuti serve, but to understand what the more run-of-the-mill spines on the penises of, say, seed beetles are for, we have to turn to Swedish researcher Göran Arnqvist of Uppsala University.

The first time I meet Arnqvist is in a small, crowded seminar room at the University of Groningen, where we are both attending a symposium on sexual selection. Slim, friendly, and sharp faced, his chin adorned with an infinitesimal goatee and his left ear with an earring, Arnqvist shows me a photo of a toilet-brush-like seed beetle penis and confides, "It looks nasty and it is nasty, too, but it is a fascinating structure!" Having first worked with water striders, flour beetles, and the occasional bedbug or bird, Arnqvist has spent much of his research time since 2002 studying the genitalia of the seed beetle *Callosobruchus maculatus.* Not that there's anything special about this species compared with the many other beetle species that have spiny penises. It's just that seed beetles are really easy to keep in the lab. Regularly furnish them with petri dishes of dried mung beans and they breed like rabbits.

In 2000, a few years before Arnqvist turned to *Callosobruchus,* Helen Crudgington, a student of Mike Siva-Jothy's (whom we met in the context of traumatic insemination), had already published an article on seed beetle genitals in the journal *Nature.* She had poured liquid nitrogen over copulating pairs, and then dissected the genitalia, frozen in action, to figure out what the spines are really for. As it turned out, the spines were certainly no harmless ticklers: they pierce the wall of the vagina and, in females that have mated several times, leave lots of tiny scars. Not surprising, then, that females try to kick away at males that keep these instruments of torture inside them for too long. When Crudgington immobilized a female's hind legs with droplets of glue so she could not kick anymore, copulation lasted twice as long as in unhandicapped females, leading to twice as many intravaginal wounds.

Crudgington and Siva-Jothy came up with two explanations for the evolution of these perforating penile bristles. Perhaps, they said, the wounds would make a female so sore that this would bolster her determination to resist subsequent males—which would be good news for the male responsible for the damage. Another, more interesting possibility was that the punctures caused by the spines would be a way for manipulative substances in the semen to enter the female's bloodstream directly.

Unable to decide between both options, Crudgington and Siva-Jothy left it up in the air—until, a few years later, Arnqvist took a stab at the problem. Literally. Together with colleagues Ted Morrow and Scott Pitnick, he carried out an experiment with several species of insects, including seed beetles, in which they caused "sublethal postmating harm" to the females—a euphemism for scientifically sanctioned insect harassment. This is what they did: As soon as a male dismounted, the researchers would hurt the female by puncturing her thorax and elytra with the tip of a very fine needle. Then they waited to see if this painful experience would make a female once bitten, twice shy about any more sexual encounters. It didn't—not in insect species with spiny penises or in species with smooth ones. If anything, the female would mate again more quickly than if she had not been hurt.

With this one hypothesis struck off, Arnqvist and his team turned their attention to the alternative explanation: that the spines would help semen substances enter the female's bloodstream more easily. Over the next few years, they (and in particular graduate student Cosima Hotzy) carried out an intricate series of experiments that proved that, like the flesh-eating substances in housefly semen, their role is to perforate the female's thick inner vagina wall, creating a passage for the male beetle's semen. They figured this out in a remarkable way. First, they fed males a radioactive diet, causing them to ejaculate radioactive semen, and after mating with nonradioactive females, the researchers could see the radioactivity seep from within her vagina into her blood. So far, so good—semen was passing into the female's blood. Next they needed to show that the spines were responsible for this. For this, they contacted Michal Polak and his laser gun.

Dr. Polak's lair at the University of Cincinnati houses an extremely accurate microlaser—the only one of its kind in the world. "It's a great system," says Arnqvist. Like a Dr. Evil of entomology, Polak can train his laser beam at a helplessly prone insect fixed in a minuscule operating theater under a microscope. The laser beam is so thin and can be maneuvered with such accuracy that Polak's machine can zap away individual hairs—and, indeed, individual penile spines. "Prft! Tzieeeeeeuw!" goes Arnqvist, in uncanny imitation of a laser beam. "It's amazing. It's absolutely amazing!" He still cannot control himself at the recollection of his student taking out individual spines, 0.05 millimeter (0.002 inch) long, on the extruded penis of an anesthetized 3-millimeter-long (0.12-inch) seed beetle.

Thus having performed microsurgery on a whole contingent of male beetles, the team managed to prove that male beetles with thirty spines removed from their penises mated just as long and ejaculated just as much semen as males with only ten spines missing, but they caused fewer wounds in the female vagina, and much less of their semen made it across the vagina wall into her blood. Finally, they also were able to show that this had an impact on the number of their mate's offspring that the males fathered: the more spines remaining, the greater a male's success at siring baby beetles. And if this weren't convincing enough, the team repeated all these experiments by comparing males born naturally with longer or shorter spines. Again, long-spined males fared better in the baby-making department.

The implications are clear: spiny penises in seed beetles probably evolved because they make it easier for the proteins in a male's semen to be thrown directly into the chemical cogs and wheels of a female's hormonal system, thereby reducing her tendency to mate with other males or in some other way hijacking her sex life to suit his interests. But this does not mean that all priapic spininess in the world evolved in the same way. And it also does not explain why early humans lost theirs.

Humans are among the smooth minority of primates. The males of most species of ape, monkey, tarsier, lemur, galago, and loris, on the other hand, have tough little spikes on their glans or sometimes all along the shaft of their penis. Even our close relative, the chimpanzee, has such roughness in the groin area. And although the spines, little nail-like outgrowths of the skin, are minute in many primate species, they can sometimes become quite impressive. In galagos, for example.

Galagos—or bush babies as they are sometimes called because of their wailing calls that emanate from the forest at night—are small nocturnal gremlin-like primates from Africa. Most of the species are tiny, weighing in at just a few hundred grams, and live much of their big-eared, large-eyed lives clutching tree trunks in the forests of the great continent. Only a few decades ago, zoologists thought there were at most seven species. Now the number stands at forty and counting. And most of those new species have been discovered not by exploring far-flung fragments of rainforest, but by intent peering between the legs of known galago species. As it turned out, many new species were hidden among more familiar species.

The male genitals of galagos proved to be a taxonomist's gold mine. While many newly discovered bush babies are pretty similar on their furry outside, their penises betray their distinctness. Skin markings, shape and size of the glans, the extent and form of the penis bone—all these are features that help distinguish one galago species from another. But prime among these characteristics are their penis spines. Most galagos have them, but some species have small ones of just a fraction of a millimeter, whereas others, such as the critically endangered Rondo dwarf galago (*Galagoides rondoensis*), discovered in 1997, has a bunch of mean backward-pointing teeth of up to 3 millimeters (0.12 inch) long. On a thin club-shaped penis of barely 2 centimeters (0.8 inch) total length, that's massive. And there is not just variation in the spine size, but also in their design. Alan Dixson, whom we met before in this chapter in his role as primate mating plug classifier, has cataloged all manner of spines on galago penises, and distinguishes three different kinds: type 1, small and short; type 2, large and thicker at the base; type 3 (hold your breath), *multipointed.*

It's a pity that we do not yet know whether these primate penile spines serve the same function as the ones in seed beetles. On the face of it, it seems likely that they, too, could rupture the skin of the inner vagina, allowing some of the semen proteins entry into the female's blood. It is known from rodents with spiny penises that after a couple of copulations, females' vaginas get so raw that they refuse any more hanky-panky.

Alan Dixson came close to answering this question. Back in 1991, he published an article in the journal *Physiology and Behavior* in which he described his penis despining experiments on male marmoset monkeys, tiny treetop primates from Central and South America. Not with a microlaser, as Arnqvist used on his seed beetles, but with simple commercial depilatory cream. Since the spines are made of keratin, the same substance that hair consists of, dehairing cream removes penile spines just as it does unwanted facial or other hair. Dixson discovered that the males that had lost their spines had more trouble finding the female's vagina after mounting her, leading the researcher to speculate that the spines may have a sensory function. Unfortunately, he did not test their effect on females, so we still do not know whether perhaps they also are important in sperm competition.

Incidentally, geneticists have figured out that, not too long ago, our immediate ancestors must also have had spines on their penises. When a team of biologists and bioinformaticians led by David Kingsley of Stanford University compared the genomes of mice, men, and chimpanzees, they discovered that humans lack a large chunk of DNA containing the switch that controls the gene responsible for keratin spines on mouse and chimpanzee penises. Presumably, somewhere along the evolutionary line between us and the common ancestor of us and chimps, this DNA got nixed. Some people believe that the wreath of soft "pearly penile papules" that about 20 percent of all men carry on the rim of their glans may be a vestige of the penile spines that our ancestors once may have had. But Kingsley thinks this unlikely. First of all, he writes on his Web site, all men, with or without papules, lack the DNA switch for spines. And second, the papules are soft lumps, not hard spines like in other primates.

The answer to the function of penile spines in other primates is up for grabs. Perhaps someday soon an intrepid primatologist with a tube of over-the-counter depilatory will take up the gauntlet—and perhaps figure out how humans got to lose them, too.

With that, we have come to the end of our evolutionary tour of the genitalia of primates, insects, and other so-called "gonochorist" animals—the term used when an individual comes as either a male or a female. We have seen that male genitals can do anything from gentle persuasion to all-out coercion, sometimes aided by sinister semen. Likewise, females' crotches are not passive sockets but rather sophisticated sorting machines that select the best of the males on offer, while at the same time trying to stay one step ahead of their mates' manipulation.

But our story is not finished yet, for not all animals that copulate and have genitals are gonochorists. Get ready to ante up: in the final chapter, we will turn to those jacks of all sexual trades, the hermaphrodites. A hermaphrodite has the tools to be both male and female, to produce eggs and sperm, and to fertilize and be fertilized. So what do we expect for the evolution of their genitalia? Wouldn't they be the epitomes of compromise? Well, quite the contrary, as we shall see.

Chapter 8

Sexual Ambivalence

The two mating *Deroceras praecox* slugs on the video screen have been circling each other for more than two minutes now, languidly licking each other with some kind of tongue. A few of the students in Joris Koene's mollusk anatomy class are beginning to go glassy-eyed and look away. This is not the spectacular slug sex they had been promised! But then Koene calmly announces, "Here it comes," and forty pairs of eyes, some in mid-yawn, see how the two gently mating slugs—suddenly, without any warning, and with un-slug-like rapidity—eject huge, bluish blobs from their genital openings, from which is thrown a delicate translucent multi-fingered organ. A collective exclamation of awe rises from the crowd of students. Precisely simultaneously, like nets cast by fishermen, the glove-like protrusions that descend over both mates' backs are then—together with the blobs to which they are attached—withdrawn almost as quickly as they appeared, and that is the end of it: two white masses are glimpsed to change owners, all genital bits are neatly tucked away in the slugs' sex openings behind their right cheeks, and the slugs part and slither their own way. "Okay," says Koene. "Who knows what they just saw?"

What the students saw, of course, is just one snippet from the exotic world of simultaneous hermaphroditic genitals. Slugs, like most other land mollusks, many flatworms and earthworms, and lots of even more obscure animal types, are male and female at the same time. They produce eggs as well as sperm, they own a penis and a vagina plus the necessary plumbing, and they apply both of them in one go when they mate and fertilize each other's eggs. (Most hermaphrodites will not resort to the ultimate narcissism of fertilizing themselves unless under duress—some species even put their own sperm in minuscule "condoms" to prevent self-impregnation.) And mutual fertilization—two-way copulation—makes for some spectacular footage, like the video produced by German slug researcher Heike Reise that Joris Koene showed to his students. Similar footage has slowly begun to leak across the divide between taxonomists and evolutionary biologists.

Taxonomists have delved into the genitalia of hermaphroditic animals with as much gusto as they have in animals with separate males and females. The finger-like protrusions on the penis of *Deroceras praecox* are well known, as they had been duly recorded in the 1960s when Polish taxonomist Andrzej Wiktor discovered and named this species. Most of the hundred or so *Deroceras* species, occurring throughout Asia, Europe, and North America, are very similar on the outside and resemble, well, slugs: thin on one end, a bit thicker in the middle, and thin again on the other end. But their penises are all different. Actually, "penis" is a bit too simple a word to describe the intricate structure, which is why *Deroceras* specialists prefer to speak of the "penial complex."

The penial complex features—of course—the penis, which looks puny at rest but when aroused can be explosively inflated to almost half the

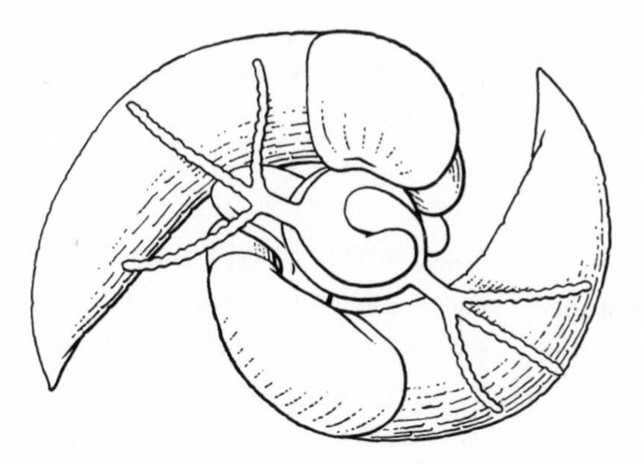

Mutual consent. The hermaphroditic slug *Deroceras praecox,* at the cusp of mutual insemination, throws a multifingered gland onto its partner's back, which probably contains hormone-like substances that increase the partner's willingness to accept its sperm.

....

slug's body length. The penis is usually not a simple schlong, but is adorned with one or more pouches and finger-like extensions. These play a role in the transfer of the slug's own sperm to the tip of its mate's penis (we will see more of this penis-to-penis copulation later in this chapter) and in transferring its mate's sperm into its own vagina—the white masses that were seen to trade places at the end of the copulation. Attached to the base of the penis is a tongue-like organ called the sarcobelum (the name is derived from the Greek term for "fleshy sting"), which is what the students saw the slugs licking each other with during courtship, in the early part of the video. At the end of the penis sits the penial gland. During courtship it engorges with a certain hormone-like compound, which is then splashed upon the mate when the penial gland is cast in the final throes of copulation.

Slug taxonomists have been using the shapes of all these constituent parts—sarcobelum, penis, penial pouches, sacs, and glands—to distinguish those otherwise identical *Deroceras* species. The penial gland, for example, can be simple or branched, it can be small or large, and the fingers can be smooth or ribbed. And the sarcobelum is even more variable, ranging from the short, stout one of the widespread *D. laeve* to large structures like the one in the Central European *D. rodnae* that pretty much resembles a human tongue and with which the slugs lap each other's bodies all over.

So what do these things tell us? First of all, that *Deroceras* copulation is curiously complex for an animal that, superficially, seems rather humdrum. Why the long licking foreplay? Why carry a tongue on your penis? Why the explosive erection, but then no penetration? Why is sperm exchanged outside of the body? What is in the substance that this multifingered organ deposits on the partner's skin? And second, the *Deroceras* penial complex apparently evolves like greased lightning: it is the only organ that really differs dramatically among a large number of closely related species. And *Deroceras* is not alone. In virtually all simultaneous hermaphrodites, be they slugs, snails, earthworms, or flatworms, exaggerated penis shapes and other genital exuberance is the rule. Of course, after seven chapters of genital extravaganza, this hardly comes as a surprise. We have seen the same in many, many kinds of gonochoristic animals (animals with separate male and female individuals, as you'll recall).

Yet that such genital profligacy should also appear in simultaneous hermaphrodites is unexpected. Although Darwin, in *The Descent of Man, and Selection in Relation to Sex,* quoted his contemporary Louis Agassiz, "Anyone who has had the opportunity to observe snail love, cannot question the seductive effects of the movements leading up to the dual embrace of hermaphrodites," he assumed hermaphrodites to be immune to unbridled sexual selection. After all, with both sexes imprisoned into one body, sexual evolution in these animals would be a matter of compromise, not of radicalism. And besides, Darwin thought them too dim-witted to see a difference between this or that mate: "[T]hey do not appear to be endowed with sufficient mental powers . . . to struggle together in rivalry, and thus to acquire secondary sexual characters," he wrote. Up until the 1970s, received wisdom remained that sexual selection was lame in the hermaphrodite world.

But Koene, an assistant professor at VU University Amsterdam and an expert on hermaphrodite sex, has seen evolutionary biology make an about-face on this, as he explains to me while feeding the freshwater snails he keeps in the basement of his lab. "We've come to realize that hermaphrodites, just like gonochorists, follow Bateman's principle and compete to fertilize a limited number of eggs," he says, stopping at a bubbling aquarium in which some twenty hermaphroditic pond snails crawl around. As he lifts the lid off the aquarium and plops in a few lettuce leaves, he explains: "Each of these snails carries some one hundred eggs, so let's say there are two thousand eggs waiting to be fertilized in this tank. If everybody mates with everybody else, then each snail receives sperm of all nineteen other snails. So any way in which a snail can tip the balance in favor of its own sperm will be evolutionarily selected."

In other words, just like in gonochorists, a particular snail that has ways to persuade fellow snails to fertilize a larger than average proportion of their eggs with its sperm will cast its genes more widely in the next generation. Over a number of generations, most snails will be this successful snail's descendants and will have inherited his/her superior persuasive or manipulative skills and devices.

Still, following Darwin's argumentation, we might expect hermaphrodites to be less amenable to such evolution, right? What with their being sexual generalists, rather than specialists? Wrong. "What Darwin was thinking of when he wrote that," explains Koene, "is things like colorful displays and ornaments. Indeed, you don't often find those in hermaphrodites." But this seems to be more than compensated for by the evolution of particularly vicious genitalia. We have already come across a fair bit of traumatic insemination and harmful genitals in gonochoristic animals. Those examples pale in comparison to what is considered normal in the hermaphrodite realm.

There are several reasons why hermaphrodite genital evolution has been taken to such extremes, according to Koene. You have to realize that for, say, a male insect to manipulate the reproductive system of a female, evolution has had to "invent" substances that latch on to key spots in the female physiology. After all, most of the genes that could produce real female hormones are switched off in a male insect. But in a hermaphrodite, the full set of male *and* female functions is turned on in all individuals. So a hermaphroditic slug could simply take some of its own female hormones and splash them onto its partner. Provided it packages the hormones properly so that they don't interfere with its own female system, nothing stands in the way of this much more efficient way of manipulation. Therefore, hermaphrodites have much easier access to the tools needed to wield manipulative genitals.

Another reason why genitalia and sexual behavior are somewhat different in hermaphrodites compared with what we are used to in animals with separate genders is that they have fewer reasons not to copulate. Whereas a female animal will mate only if she stands to gain something—usually meaning better fathers for her offspring—a hermaphrodite always has two options to weigh: it may not need to receive any more sperm for its eggs, but it would still be interested in donating some of its own sperm to the gene pool. The result of such split personalities is that when two hermaphrodites meet, there are four reproductive decisions to take, four roles to play, rather than just two, as is the case in male-female encounters. This means that in the complex maze of sexual opportunities that present themselves to a roaming hermaphrodite, a biochemical "yes!" will form in its brain much more frequently than in gonochorists. So, paradoxically perhaps, having done away with males and females has opened a portal to a more sexual, more promiscuous universe in which evolution has greater, rather than reduced, opportunity for experimentation.

And experiment hermaphrodites have. Over the past fifteen years or so, evolutionary biologists have woken up to the fact that hermaphrodites live in a world where bizarre and often gruesome mating rituals involving particularly outlandish sex organs are the norm. Take, for example, the penis fencing in *Pseudoceros bifurcus,* a hermaphroditic marine flatworm of the Indian and Pacific Oceans. Practicing hit-and-run traumatic insemination with paired dagger-like penises, these worms, upon meeting a potential partner, generally prefer to inseminate rather than to be inseminated. So they rear up and engage in a duel of striking and parrying. An even better example is the love dart of land snails. The function of the calcareous arrow that many snails "fire" at close range from a special organ at the base of the penis had remained a mystery until the late 1990s. For a long time, it was thought to be a kind of courtship—a dowry in the shape of a bit of packaged chalk to be considered in cryptic "female" choice. But, as we shall learn in the next section, the truth is less pacific.

Evildoers, Evil Dreaders

Naturalists have marveled at snail sex and the associated wielding of love darts since the days of Aristotle, and one of the nicest descriptions comes from early nineteenth-century British zoologist Thomas Rymer Jones. "The manner in which [snails] copulate is not a little curious," Jones writes in his *General Outline of the Organisation of the Animal Kingdom.* "After sundry caresses between the parties, during which they exhibit an animation quite foreign to them at other times, one of the snails unfolds from the right side of its neck . . . a sharp dagger-like spiculum or dart. . . . Having bared this singular weapon, it endeavours, if possible, to strike it into some exposed part of the body of its paramour, who, on the other hand, uses every precaution to avoid the blow, by speedily retreating into its shell. But, at length having received the love-inspiring wound, the smitten snail prepares to retaliate, and in turn uses every effort to puncture its assailant in a similar manner." He then goes on to describe the "more effective advances" (penis erection and mutual insemination) that ensue, but closes with a sentence that drily encapsulates the wonder of the naturalist who thought he'd seen everything: "Such is the peculiar manner in which the amours of snails are conducted."

Peculiar indeed. In the edible escargot *Cornu aspersum,* during mating both snails launch a loose, nearly 1-centimeter-long (0.4-inch) limestone needle into each other's flank. (At least, that's where it's supposed to go; it sometimes misses its target and ends up on the floor—or, worse, impales the partner's head.) Next time you eat snails, you'll know what it is that's crunching between your teeth. But love darts are not unique to escargots, although this is probably the species in which Jones observed them. Similar weaponry is employed by many families of land snails and also by sea slugs and earthworms. In Europe, all Helicidae (the family to which *Cornu aspersum* belongs) have them, and also Chinese camaenid snails, the colorful *Polymita* from Cuba, and the ariophantid snails and slugs that live all over Southeast Asia.

And although all are fired from the same sort of "dart sac," a muscular bag that sits at the base of the penis, love darts differ quite a bit between snail species. In the Eastern European *Monachoides vicinus,* they look exactly like medieval archers' arrows, complete with head, shaft, and fletchings. In *Everettia corrugata* from Mount Kinabalu on Borneo, they are hollow and furnished with a row of holes along the side, and in the Italian *Marmorana* snails they are, quite fittingly, the spitting image of a Roman soldier's sword.

Not only do the darts look different, they are also differently employed. Whereas *Cornu aspersum* carries a single dart that it shoots once, then growing a new one, the African slug *Trichotoxon heynemanni* has two darts, which it uses simultaneously, and in the Japanese "samurai snail" *Euhadra subnimbosa* mates stab each other a staggering thirty-three hundred times during a single bout of courtship. Some snail species always shoot a dart; others do so only with certain partners. Some shoot before penetrating each other with their penises, others during, and yet others after. With such a diverse array of darts and dart-wielding behavior throughout the world of snails and slugs, there must be some important advantage to it, since, in evolution, elaborate organs and behavior will be done away with if they are useless.

For years, biologists had speculated that that advantage might be a "nuptial gift" of calcium. Snails build their eggshells, like the shells they themselves inhabit, from calcium carbonate, which can be hard to come by in some habitats. So if you supplement mating with a generous gift of

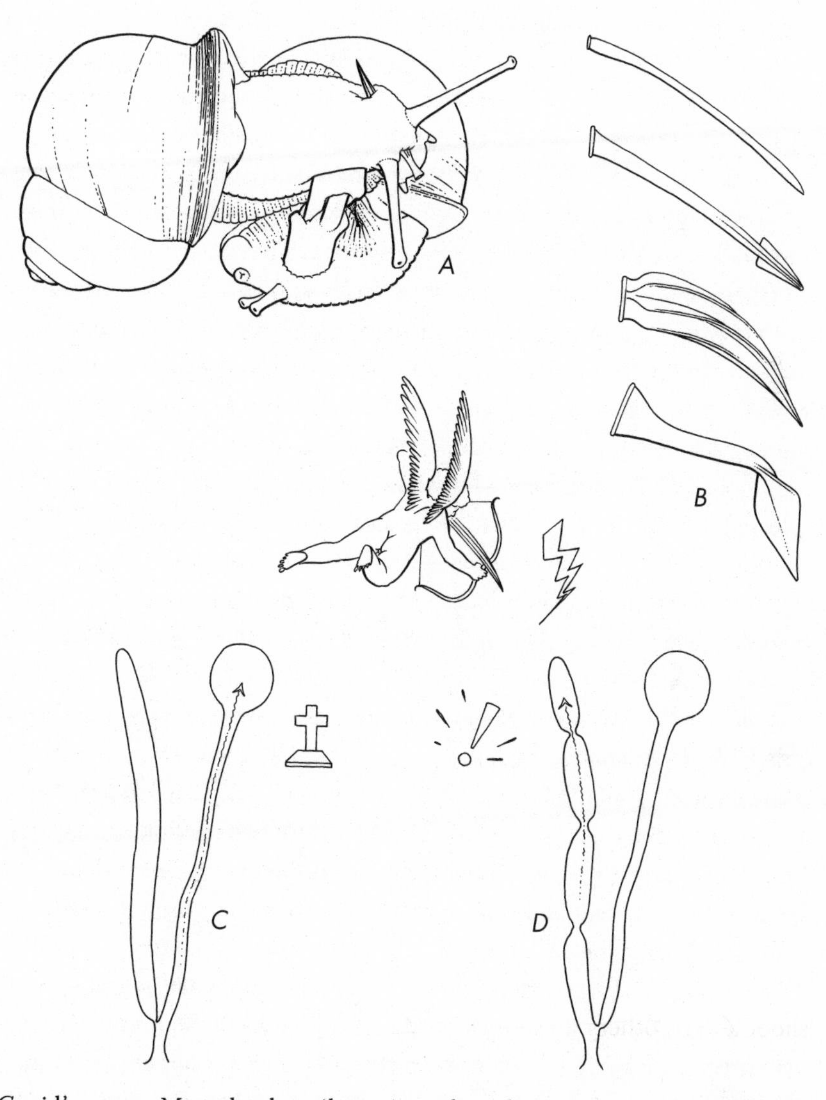

Cupid's arrow. Many land snail species, when they mate, impale each other with a "love dart" (A). In different species the darts have different shapes (B). Without dart shooting, almost all sperm is broken down in the balloon-shaped bursa of the recipient (C). The dart is laced with a hormone-like substance that closes off the bursa and keeps the sperm in the "safe" diverticulum (D).

....

calcium to your partner, conveniently deposited hypodermically, your partner may be able to lay more eggs, gratefully fertilized with your sperm.

It was such a plausible hypothesis that it went untested for decades, until, in the 1990s, Canadian biologist Ronald Chase of McGill University in Montreal decided on a proper check of the idea. To do so, he hired Koene, then a young graduate student. "We were skeptical of the nuptial gift notion," Koene explains. "Nuptial gifts may serve a purpose if they are given by a male to a female, but in simultaneous hermaphrodites, it doesn't follow: I give you a box of chocolates, you give me a box of chocolates. What would be the point of that?" Together Chase and Koene started a lab colony of *Cornu aspersum* snails and began measuring the quantity of calcium in the dart, the amount of dart material taken up by the "victim," and the amount of calcium needed by the maturing eggs. The results, as they reported in 1998 in the *Journal of Molluscan Studies*, simply did not add up: only one in twenty snails internalized a dart—the rest eventually squeezed them out of their skin. And on top of that, a single dart contained barely enough calcium to supply material for one egg. It was the death knell for the nuptial gift hypothesis.

Later that same year, in the *Journal of Experimental Biology*, Koene and Chase revealed what must be the true evolutionary reason behind dart shooting: sexual manipulation. They had noticed that when shot, the dart is covered with a thick mucus, produced by glands that sit on the dart sac. They then surgically removed the female reproductive organs and, keeping these alive in a petri dish, applied this mucus to them. Immediately, peristaltic movements began to ripple through the tube-like organs.

To understand what this means, we must explain what happens in the boy bits and girl bits of a snail during hermaphroditic sex. As copulation starts, the male genitalia begin packaging a large amount of sperm into a parcel, the spermatophore. This parcel is then placed via the penis into the so-called diverticulum (a blind side pocket) of the partner's female genitalia. From there, the spermatophore is eventually transferred to another organ, the bursa, which is a kind of stomach dedicated to digesting the spermatophore, sperm and all. So how do snail eggs get fertilized if all the sperm is consumed? Well, at some point between deposition in the diverticulum and transfer to its nemesis, the bursa, a few sperm cells (usually just a few hundred out of the more than a million that sit inside the spermatophore) gingerly escape via the tail of the spermatophore and travel up the oviduct toward the eggs. The more sperm escape, the more eggs may be fertilized. And this is where the love dart comes in: the mucus that the dart is laced with contains a hormone-like substance that, when the dart is shot deep into the snail's skin, triggers those involuntary ripples of the diverticulum. These help to absorb the spermatophore more quickly; they also close off the entrance to the bursa and thus increase the shooter's chances of fatherhood.

That these chances are indeed improved was proven a few years later by Chase and his student Katrina Blanchard in a beautifully simple experiment they published in the *Proceedings of the Royal Society B.* They had managed to perform microsurgery on snails—yes, this meant anesthetize them, make an incision, amputate the dart sac, sew the patients up again, and let them recover. When they then let these dartless snails mate with other snails, and injected the partners using a syringe filled with either dart sac extract or innocuous saline solution, they found that the saline had no effect, but the dart extract boosted the number of baby snails fathered by the dartless snails to twice what it would otherwise have been.

It is not unimaginable, says Koene, that whatever substance it is that *Deroceras* slugs deposit on their partners' skin at the end of mating serves a similar function as the hormone-like substance that the snail love dart is spiked with. Still, why is there so much diversity in the parts of the genitals that are involved in chemical persuasion? The biggest differences among different *Deroceras* species are in the sarcobelum, and we have seen that the molluscan Cupid has likewise provided thousands of snail species with all manner of love's arrow: long, short, thick, thin, curved, straight, round, or flanged. So why this diversity? Wouldn't we expect that all species would eventually converge on the kind of device that works best? Well, yes, we would—if simultaneous hermaphrodites weren't such sexual schizophrenics.

The male portion of a snail's or slug's genitals—its entire apparatus for packaging and delivering sperm together with the chemical cocktail injected into the partner to boost the sperm's chances—is optimized by generations of success in acting like a male. But at the same time, its female genitals—the part of the body that receives and processes spermatophores—is designed to serve its female interests best. For a healthy egg clutch it may be much better to fertilize only a few of your eggs with sperm from each mate. And digesting most of a mate's spermatophore might be one way of achieving that aim.

The evolutionary result of this conflict of interest is not a status quo. It is the same kind of never-ending dance that we have seen in the mutual accommodation and evasion of the genitalia of males and females. Here, however, it has its impact not on separate sexes but on the separate parts of the genitalia of a single hermaphroditic animal. Improvements to the male system, such as transferring more hormone-like substance by adding flanges to the dart or multiplying the number of darts, will meet a response in countermeasures in the female system, like making the diverticulum deeper and thus harder for sperm to escape from. This will then be followed by a new round of male measures (perhaps stabbing repeatedly rather than just once, and loading the dart with fresh mucus each time it is withdrawn), and yet again female countermeasures (maybe adding side pockets to the diverticulum). And since these histories of making love and war are basically unpredictable and can go in different directions and involve different "inventions" in different species, diversity is the result.

In the end, every snail or slug is saddled with the outcome of this generations-old conflict of interest: its male genitals try to do exactly what its female genitals won't allow. Perhaps one of the reasons that simultaneous hermaphrodites have been eschewed by scientists studying sexual evolution is that it is so hard for us—gonochorists as we are—to learn to think like a hermaphrodite. But once you do, says Koene, the insights gained are not to be sniffed at. "Having two sexes in one animal makes hermaphrodites twice as interesting!" he proclaims, and drops another lettuce leaf in the snail aquarium.

Castration Anxiety and Penis Envy

I once watched an Irish mollusk researcher, in a fit of slug chauvinism, try to convince an audience of biologists that slugs are superior lab rats. "Behaviorally speaking, a slug is basically a rat," he told them. "Cover a rat in slime, amputate its legs, pull its genitalia up behind its right ear, and film it in slow motion, and you've got a slug!" Where their genitalia and sexual behavior are concerned, slugs indeed are a gold mine for research with seams even richer than *Rattus norvegicus*. We have already seen what amazing tricks a single slug genus has up its sleeve. But *Deroceras* is just the tip of the iceberg. Malacologists (this is what mollusk aficionados call themselves) are slowly beginning to dish out the slug sex smorgasbord that lies hidden under rocks and stones, both on land and in the sea. And, as we shall see, this is throwing up yet more surprises about how genital evolution works in simultaneous hermaphrodites.

Sea slugs are perhaps best known for their wonderful psychedelic colors, especially in one particularly prominent and species-rich group, the so-called nudibranchs. Nudibranchs—"nudis" to friends—are named after the gills (branchia) that many species carry uncovered ("nude") on their backs. One of the more modestly colored nudis is *Aeolidiella glauca,* a species that inhabits the cold shallow-water eelgrass beds along the shores of Northern and Western Europe. They prey on small sea anemones and they carry frills along their flanks in which they store the anemones' venomous nettle cells, appropriating them for their own defense. But that's neither here nor there. I mention *A. glauca* because it is one of the few nudibranchs—out of the more than three thousand species that exist—that have had their sexual behavior properly studied.

We can credit our knowledge of *A. glauca* to Anna Karlsson of Uppsala University and Martin Haase of the University of Greifswald. Back in the late 1990s, Haase explains as I chat with him in the office he shares with his wife, ornithologist Angela Schmitz-Ornés, his boss ran into Karlsson at a conference and invited her to join forces with Haase. She was a novice Ph.D. student of reproductive biology in this common marine nudi, he a skilled anatomist of mollusks willing to try his hand at something new. Together they embarked on a series of discoveries that had their jaws dropping. And dropping. "Everything the church forbids is present in this species!" Haase exclaims.

Aeolidiella glauca sex goes like this. When a randy individual runs into a potential partner, it first starts tailgating it. Sooner or later, the object of its love will turn around and face its suitor. Within a minute, this face-to-face phase then merges into "sidling"—the pair sliding past each other's right-hand flank, where the genital openings sit. If necessary, the most eager of the two will nibble at the other's genital opening, as if to arouse the partner. First anchoring one with the other using flaps around the genital opening, the slugs then produce their huge penises. To Haase's and

Karlsson's surprise, the slugs don't insert these into each other as most other sea slugs do—the organs are too large for that. Instead, they release a sausage-shaped spermatophore from their penis, which they carefully place on each other's back. Then they lift their penises and tap down on the spermatophore several times, as if to hammer it firmly in place. Usually both slugs accomplish all this in perfect synchrony.

This phase of sperm transfer lasts only a minute or two, after which the two slither their separate ways. The real magic starts only once the mate has left. Within a few minutes, a slug with a spermatophore on its back will crane its head over its right shoulder to reach and then eat part of the spermatophore, increasing the distance between the genital opening and the spermatophore. Then it waits. Some three hours later, a thin file of sperm is seen to begin to break free from the spermatophore and to start marching over the slug's skin toward its genital opening. The sperm being tiny, it takes several hours for them to reach their destination, despite the fact that the column takes shortcuts by swerving between the nettle-cell-containing lobes on the slug's back.

The slug is not a passive bystander to the sperm parade on its back. It seems to monitor the progress closely, and when it decides that enough sperm have entered its genital opening it will interrupt the flow by sucking up the sperm from its back. "You can really watch that!" Haase says. Then, its stomach full of hapless sperm, it will begin laying eggs, using the sperm that it did not eat up to fertilize them. Haase goes on: "I even watched two animals that interrupted the sperm with their mouths, then started to lay a batch of eggs, and when that was finished resumed sperm uptake from the thread that was still traveling across their back." Haase winks at his wife at the desk opposite him. "When Angela first met me," he says, "she googled my name and to her consternation found all these *Aeolidiella* publications of mine."

Sea slug sex is sure to come up with more surprises, Haase thinks. Only in 2012 were details revealed of copulation in another nudi, the Pacific species *Chromodoris reticulata*. Japanese researchers discovered that, though this species copulates in a more conventional way (penis in vagina), after mating the penis drops off, only to be replaced by a spare length of penis that the animal keeps coiled up inside the body. The coil is long enough for three matings before the whole system needs to be regrown.

Nobody understands yet why such a throwaway penis evolved, but since sperm were found snagged by the barbs at the end of the penis, it may be a way to remove and then discard sperm from previous mates—a disposable hermaphrodite version of the sperm-scooper we have seen in damselflies.

Land slugs also sometimes lose their penis. There is a report from Kazakhstan of a species of *Deroceras* (yet another one) that occasionally, at the end of copulation, bites off its own penis and then presents it to its partner as a meal. And the large yellow "banana slugs" (*Ariolimax*) of California do it the other way around: they sometimes chew off their partner's penis at the end of coitus, sending it (now her) packing in emasculated state.

Slug salacity reaches its pinnacle in the large, up to 20-centimeter-long (8-inch) "tiger slugs" of the genus *Limax.* These animals are native to Europe, but one species, *Limax maximus,* was accidentally introduced into the Americas, Australia, South Africa, and many other temperate and subtropical parts of the world—hence, they will be familiar to many. The animals are nocturnal, slithering around under cover of darkness, leaving thick shiny mucus trails on the ground and on trees they climb. But even by day they are hard to miss: with their pantherine, pied sandy-and-black presence they take up a lot of psychological space whenever one inspects underneath a log or stone or when pulling away a piece of dead tree bark. Though *Limax maximus* is the best known and the most widespread, there are in fact dozens of species, ranging in color from cream via deep red to sooty black, often with darker spots and stripes. And many of these lesser-known *Limax* species inhabit small areas in the Alps and other mountainous parts of Europe.

Still, the diversity of *Limax* would probably have long remained unknown and unstudied if it weren't for their peculiar genitalia. Remember barnacles? In Chapter 3, the burrowing barnacle *Cryptophialus minutus* was hailed as the animal with the world's longest penis relative to its body size—eight times longer. As it turns out, some species of *Limax* are serious contenders for this title—a reputation that seems to speak to the imagination of some people. So much so, in fact, that a veritable scientific dynasty of respectable gentlemen-naturalists, three and a half centuries' worth of them, have crossed swords over the nitty-gritty details of courtship, copulation, and—especially—penis length in *Limax.*

The debate was opened in 1678 by English naturalist Martin Lister. In his treatise on English animals, *Historia Animalium Angliae,* he described, in Latin, how he had seen two large spotted slugs lower themselves in passionate embrace along a foot-long slime thread from a tree trunk while extruding and entangling their large, pale bluish penises into a bulky pear-shaped knot. A few years later, in 1684, his Italian colleague Francesco Redi (more famous for disproving the old belief that maggots are spontaneously generated in rotting meat) published a book containing his own observations on the mating of *Limax.* They could not be more different from Lister's account. Redi did not mention any long slime threads, but instead spoke of entwined slugs, dangling down in apposition their extremely long penises, longer than a "florentine yard"—about 75 centimeters (30 inches)! He even provided a plate showing two rather cheerful-looking spotted slugs in possession of extremely long, intertwined penises that extended to the bottom of the page, far enough for Redi to use the rest of the space for other detailed drawings of slug anatomy.

In the following centuries, a whole procession of natural history authors added their penny's worth on *Limax* coitus. And time and time again, the observations confirmed Lister's account, not Redi's. Some small discrepancies were resolved by figuring out that there were, in fact, two species: *Limax maximus* descending from a long slime thread and engaging in a tight penile handshake in the form of flanged knot; and *Limax cinereoniger,* hanging by its tail tips and entwining the penises to form a dense bell-shaped coil. But while the penis sizes of these two species were respectable, they never approached the lengths claimed by Redi. So, with the advancing knowledge of *Limax* coitus, Redi's observations began to sound more and more ridiculous, and by the early twentieth century H. Wallis Kew dismissed them as "quite obscure." Surely, Redi must have exaggerated the priapic prowess of his Latin slugs! Either that or he had mistaken slime threads for penises.

But in the first few decades of the twentieth century, as northern European malacologists leaned back, secure in the knowledge that all *Limax*'s mysteries had been solved, isolated reports began to trickle in from Italian and Swiss authors that seemed to confirm that Mediterranean *Limax* were indeed unbelievably well hung. Confusion reigned, but then came 1933. In that year, German zoologist Ulrich Gerhardt of the University of Halle published the first of a series of groundbreaking papers on slug sex. Apparently adopting a broad interpretation of his mandate as staff scientist at the Institute for Anatomy and Physiology of Domestic Animals, he carried out meticulous studies of slug behavior in large indoor cages. He did dissections, timed and filmed copulations, and, in good German tradition, described his discoveries with painstaking detail, here and there quasi-respectfully lashing out at previous, less patient colleagues. Having thoroughly documented the mating in *Limax maximus* and *L. cinereoniger* in more detail than ever before, he turned to the problem of Redi's generously endowed *Limax*.

Bernhard Peyer, one of the Swiss biologists who had, a few years earlier, claimed to have seen the same as Redi, mailed fifty live Swiss slugs to Gerhardt, who housed them in tall cages (tall enough to accommodate dangling Florentine yards—just in case). One evening, about two weeks after the animals had taken residence in their new accommodation, one slug began showing interest in another by persistently following the slime trail of the object of its desire. Two hours later, by 10:00 p.m., the pair was circling each other on one of the vertical slats supporting the cage, creating a puddle of slime in the center. Approaching each other closer and closer, they began to entwine and started to release their heads, then the rest of their bodies from the surface, until they eventually hung by their tails from the sticky blob of slime. Then, ever so slowly, their penises began to peek out from their genital openings.

Gerhardt waited up, notebook at the ready. By 11:30 p.m., the penises were still only 2 centimeters (0.8 inch) long, and had just begun to touch and stick together. Mating in *Limax* normally lasts only an hour or so, so one can imagine that Gerhardt at that moment must have had his doubts as to whether those long slug schlongs were ever going to form. But he waited, and waited, probably with the aid of many cups of coffee, and through the night both penises steadily grew under a slow pumping motion that let the tips swell and then shrink again in lively dance-like motions. At 3:45 a.m., they were 8 centimeters (3 inches) long. By 7:00 in the morning, 26 centimeters (10 inches), and Gerhardt could record in his notes that rice-grain-sized spermatophores were seen to be descending in the thin, bluish-translucent penises. At 10:00 the next morning, a full twelve hours after mating had begun, the penises reached their maximum length: the combed tips clung to each other a whopping 80 centimeters (32 inches) below their owners, precisely as Redi had said 250 years earlier. Since the slugs themselves are some 13 centimeters (5 inches) long, this comes close to the record-setting barnacle penis.

In this extended state, the pair hung still, meanwhile squeezing their spermatophores to the very ends of their penises. As soon as the spermatophores of both partners reached the end of the line, both slugs suddenly began hoisting in their penises while at the same time expelling the spermatophores from the orifice at the end of the penis. Then, with a sleight of hand so rapid that Gerhardt could dissect the events only by filming them at sixty-four frames per second, the comb-shaped penis tip unfolded to form a scoop-like tool, which first caught its own spermatophore, stuck this to the scoop of the partner, and simultaneously received the partner's sperm package. Both animals then withdrew their penises with lightning speed, meanwhile sucking in the foreign spermatophore through the penis's opening. By 11:00 a.m. the couple began to separate; half an hour later the final bits of penis were tucked away inside their respective genital openings, and both partners, hopefully as satisfied with the outcome as Gerhardt was, slithered away.

The genitals and the behavior of this *Limax* proved so exceptional that Gerhardt considered

Sexy slugs. *Limax* slugs engage in exchanging sperm via their lengthy, dangling penises, which, in these hermaphrodites, are used to give and also to receive sperm.

the Swiss slug a separate species, which he named *Limax redii* to honor the discoverer of their unique mating behavior. Like many a lab rat, the two specimens that allowed Gerhardt his first peek into their love life then donated their bodies to science: they float, paled by almost eighty years of immersion in spirit, with pickled penises, in two glass flasks on a shelf in the natural history museum in Berlin.

In recent years, *Limax* has become a popular object of study among amateur and professional biologists in Austria, Switzerland, France, and Italy. Several Web sites are devoted to their biodiversity, which has led to the discovery of many new species. Rather than the handful known in Gerhardt's day, the count now stands at thirty-three, and several of those—all occurring in Italy, Switzerland, and France—are now known to have similarly long penises as *L. redii.* On the Italian Web forum NaturaMediterraneo.com, latter-day disciples of Redi excitedly report their discoveries. Photos show proud malacologists standing by trees from which mating *Limax* hang, holding tape measures to document the record-breaking lengths of slime threads and penises. In one, of a *Limax* pair from Corsica, penis length reaches a calibrated 92 centimeters (37 inches). "Nuovo record del mondo," the caption reads.

Besides confronting us with the unsettling notion that a penis can be both a sperm-delivering and a sperm-receiving organ in a hermaphrodite, the exaggerated *Limax* phallus is a good example of one of the main differences of opinion in the field of genital evolutionary biology. On the one hand, it could be the result of cryptic female choice. Or rather, "cryptic choice," since we are talking hermaphrodites here. If longer penises are a preferred ornament, like a peacock's tail, they would grow longer and longer during evolution. However, it could also be a sign of sexual manipulation: a struggle over who ultimately gets to play a more male-like role. The exact mechanics of the prestidigitation that goes on at the end of *Limax* copulation are not fully known yet, but perhaps a slug with a slightly longer penis is able to donate its spermatophore while at the same time avoiding accepting the partner's.

And the remarkable copulation of *Aeolidiella glauca* may have evolved in a similar way. Other nudibranchs place their spermatophore closer to the genital opening, but there the receiving slug may easily remove it if the receiver feels so inclined. By placing the spermatophore out of reach on the partner's back, and having its sperm sneak into the genital opening stealthily when the host isn't looking, the ancestors of *A. glauca* would have had an advantage, at least until sperm sucking evolved. Koene admits that, faced with such bizarre hermaphrodite sex, he tends to think of arms races rather than of cryptic choice. "They have manipulation written all over," he says. But then he adds that it is very hard to disentangle the two. "After all, even slugs may like to mate with manipulative partners," he concedes.

Left Right After Sex

In this last section, I will showcase some of my own research in genital evolution to add yet another, final layer of complexity. After thoroughly immersing you in the baffling ins and outs of sexual selection acting on the genitalia of animals that come in male and female versions, I then asked of you to wrap your head around the confusing dualism of hermaphrodite reproduction. In this final section of the book, we will journey briefly into the strange looking-glass world of mirror-image hermaphrodite genitals, where everything is the same and yet completely different. At the end, we'll come to see that, under certain conditions, genital evolution can even influence the shape of the entire body of an animal.

The story begins in 1989, when I was a young biology student arriving in Kuala Lumpur, Malaysia, to spend five months studying the particulars of a minuscule parasitic wasp called *Trichogramma* as it searched for the moth eggs that its larvae grow up in. (It was actually somewhat less frivolous than that may sound: the wasp was to be used to help control the caterpillars that were destroying crops of sweet corn.) But before starting my project—which turned out to involve mostly the staking out of fields of corn, inspecting the growing stalks for caterpillar frass while chasing off rampaging water buffaloes—I went on a short vacation to the sandy, rural, peaceful east coast state of Terengganu. On the recommendation of a local fisherman, I took a boat to a small offshore island named Kapas.

Today, Kapas is a bustling tourist destination where a dozen beach holiday operators ply their trade for boatloads of local and foreign visitors. Back then, it was deserted, save for six A-frame chalets rented out by a stray Englishwoman. The island's two kilometers of pristine coconut-palm-fringed beach were the sole territory of the handful of chalet guests, the occasional green turtle paddling up the beach, and thousands of ghost crabs. It was my first time in the Tropics, and I could not get enough of trekking along the shore and through the forest and the coastal scrub. I saw my first hornbills there, my first flying lizard, and my first *Nephila* spiders—hand-sized frights that spin gigantic golden gossamer orb webs. But even more memorable than those were the *Amphidromus* snails suspended from trees and branches all over the island.

I knew about *Amphidromus* because the year prior to my Malaysian adventure I had spent in the malacology department of Leiden's natural history museum, where my adviser, Edmund Gittenberger, had taught me that the snails of this genus—large, colorful tree snails occurring everywhere in Southeast Asia—are the only ones in the world that come in what appear to be even mixes of clockwise- and counterclockwise-coiled individuals (in fact, *Amphidromus* means "turning both ways" in Greek). Almost all other snails, all 150,000 species of them, are either clockwise or counterclockwise, but never both.

The reason why snails normally don't come in both types is easy to understand if we read an article written a century ago by the German naturalist Johannes Meisenheimer. Meisenheimer had found a contrarily coiled Burgundy snail (*Helix pomatia*). Normally, this snail twists clockwise, meaning that when one holds its shell with the tip up, the aperture (the opening from which the animal emerges) would be facing the viewer on the right-hand side. But the mutant that Meisenheimer had found was a mirror image, coiling counterclockwise, so with the aperture on the left. Such mutants occur now and then in most land snails, but they are extremely rare—often just one in ten thousand, or even rarer.

When Meisenheimer put his mutant in a jar with a regularly spiraling animal, it was clear that, at least in snails, being each other's opposite does not make for a happy marriage. After all, it is not just the shell that is reversed; the animal's whole body is also back to front. Its genital opening is on the left side of its head, not its right; its dart shoots leftward, not rightward. Its entire genitalia are mirror-imaged, and so is its brain—even its sexual behavior. The result was that "for days and weeks the animals fatigued each other in courtship, without ever achieving a final copulation," Meisenheimer observed with some sympathy. This means that a left-handed snail, if it should appear smack in the middle of a population of right-handers, would be dead on arrival, evolutionarily speaking: it would not have anybody to mate with. That is why all snail species are either clockwise *or* counterclockwise. And that is why *Amphidromus* snails, flouting this rule so openly and brazenly, are so special.

I dug up from long-term storage my memory of the *Amphidromus* of Kapas when, many years later, I landed a position in a Malaysian university and began a search for interesting research projects. The island of Kapas, with its particularly high density of *Amphidromus* snails (not to mention the idyllic setting), seemed the perfect place to solve that still puzzling conundrum of left- and right-coiled snails living side by side. So with a motley crew of colleagues, students, and volunteers, I set up a research project on Kapas (no longer pristine, but still very pretty) to unveil the secret of its mollusk inhabitants.

We had two study sites. Site 1 was in a bit of forest behind the Lighthouse Inn, a ramshackle hostel run by Din, a magnanimous, lanky, long-haired Malay with Rastafari leanings. Site 2 was farther north, in trees just behind the only beach that was still quiet enough for turtles occasionally to land there to bury their eggs (only to have them consumed immediately by the huge, dinosauresque water monitor lizards). In both places, the snails were plentiful and clearly came in two mirror-image versions: about 35 percent clockwise coiled, the rest counterclockwise coiled. We set about interrogating them in the only way that field biologists know how: by getting our hands dirty and scrutinizing their lives up close in the trees.

Enveloped in clouds of mosquitoes and against the background reggae music wafting in from Din's place, we first checked if, like in all other snails, the coiling direction is genetic in *Amphidromus.* It was. Baby snails take the coiling direction that is dictated by their mother's genes, so all young snails hatching from a nest of eggs should coil in the same direction. And that is precisely what we saw in the jelly-pearl-like egg clutches that the snails deposit in wet crevices in rotten branches and stumps: a family of newborn snails were either all clockwise or all counterclockwise.

Next, we figured, perhaps the lefty and righty snails have different roles in the ecosystem. Did they eat different things, live in different parts of the vegetation, or were they themselves being eaten by different predators? Nope. We spent a year writing numbers on the shells of live snails and following them around, but whatever their bent, the snails all mixed with alacrity, feeding and resting side by side in the same places. And the island's forest rats that bite off the top whorls of the shell and then suck out the flesh like true gourmands also did not seem to care between left and right. They cracked and ate both in the exact proportions at which they occurred.

But while clambering around the foliage tracking marked snails, we began to notice something peculiar: now and again, on wet days, we found a left-coiled and a right-coiled snail having it off. Unlike Meisenheimer's snails, they did not "fatigue each other for days on end" without ever consummating their desires. We began paying more attention to copulating snails. This was not easy, because the snails are not of a very passionate demeanor and mate only occasionally, usually during the onset of the monsoon season. Still, over the course of several years, we caught more than a hundred pairs of snails with their trousers down, as it were, and a surprisingly large number of those—in fact, more than expected even if they paired off randomly—were of a clockwise copulating with a counterclockwise. Apparently these mixed couples did not mind the fact that all their organs were in the wrong places, or perhaps even *preferred* to mate on the other side of the mirror-image divide.

One evening, as the thunderclouds were building up rapidly over the Terengganu coast, three of us sat down on a fallen coconut stem to work in assembly line style on marking a large pile of live snails with permanent marker, covering the marks with clear nail varnish, and writing down the details. Meanwhile we discussed the problematic sex lives of our snails. Ignoring the stunned stares of passing vacationers in swimming gear, we decided that we needed to find out what went on inside a mating pair of snails. How and, more important, *why* would *Amphidromus* seek out mates of opposite coil while all other snail species don't and won't?

Do you recall the *Silhouettella* spiders whose genitalia Matthias Burger studied, back in Chapter 3? And Helen Crudgington's study of mating in seed beetles (Chapter 7)? They could discern the exact internal workings of the genitalia of these animals by pouring liquid nitrogen over mating pairs. As against my malacophile grain as this would be, we had no option but to replicate the same dreadful deed on these mating *Amphidromus* snails. Since bringing a gallon of liquid nitrogen to Kapas was not feasible, I brought the next best thing: freezer spray. Applied as a local anesthesia and also for cooling electrical components, a volatile liquid that freezes whatever you spray it on would also do the trick. So one year, as the snails began to respond to the first monsoon rains with their amorous escapades, I hiked around the island catching pairs in the act and booking them for eternity. After a few deft sprays, each couple, joined forever, sank to the bottom of a flask of alcohol—even death did not part them.

Back in the lab, I dissected the copulas. Mating in *Amphidromus* is not a fleeting affair: it lasts from dawn till dusk, and my specimens showed all the stages in this drawn-out process. As soon as mating takes off, both partners begin producing a 6-centimeter-long (2.5-inch) looped spermatophore. (Remember, we're still talking simultaneous hermaphrodites here, so they fertilize each other at the same time.) They build this sperm package in the epiphallus, an extension to the penis, the inside of which functions as the mold in which the spermatophore is cast. As it turned out, the very tip, where the sperm eventually leave the package, was produced in a corkscrew-shaped appendix of the epiphallus. And what do you know? The coiling direction of the corkscrew turned out to be the same as the coiling direction of the snail itself: a clockwise snail has a clockwise corkscrew at its spermatophore tip, a counterclockwise snail a counterclockwise one! This feature is rather unique to *Amphidromus;* most other snail species have straight spermatophore tails.

This was beginning to get interesting. Now let's move to the female side: the recipient. There, the spermatophore gets crammed in the bursa, that organ whose task it is to digest most of the spermatophore and all that is in it. Only that corkscrew tip sticks out of the bursa into the spot where the tube leading up to the eggs attaches. And this is where the magic happens. When I cut open the female genitalia, I saw that this tube attaches to the bursa under an angle; this angle is to the right in a clockwise snail, to the left in a counterclockwise snail. As I was quietly sitting on my lab stool behind my microscope, dissecting away, it began to dawn on me: the upshot of all this mirrored and contorted genital plumbing was that the mesh between a spermatophore tip and the female genitals *is better if the mates are of opposite coiling direction!* And this would probably also mean that more sperm made it out of the spermatophore and into the tube heading for the eggs! I made a little pirouette on my lab stool when I finally figured this out.

The situation in *Amphidromus* perhaps reminds you of the duck genitalia of Chapter 6 that show the reverse: a counterclockwise penis does *not*

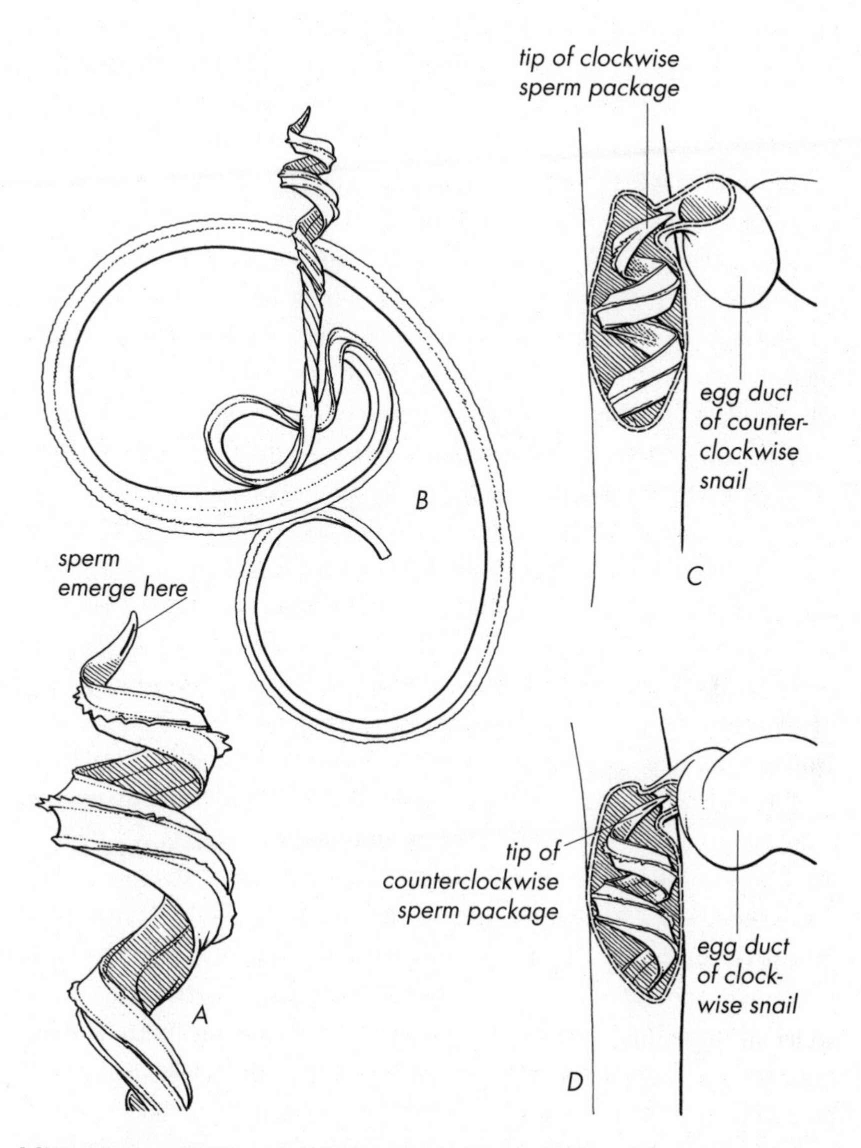

Mirror-image mating. *Amphidromus inversus* snails come in two forms: coiled clockwise and counterclockwise. This applies not only to the coil of the shell but also to the tip (A) of the elaborate sperm package (B). As it turns out, a clockwise sperm package fits better in the female reproductive system of a counterclockwise animal (C) and vice versa (D).

....

fit properly into a clockwise vagina. The funny thing is that fifty-fifty mixtures of clockwise and counterclockwise vaginas will never evolve in ducks, because the female will not benefit from making it easier for a male to penetrate her. In a simultaneous hermaphrodite like *Amphidromus,* things are different, because both partners, since they are both acting as males, benefit from mating with a countercoiled partner.

Paul Craze of the University of Sussex (and current editor of the esteemed journal *Trends in Ecology and Evolution*) came to Malaysia to work on *Amphidromus* with me, and he wrote a computer program to simulate what the effect of the snails' hand-in-glove-like genitalia would be. Sure enough, Paul's computer simulations showed that this would indeed produce a snail species with both coiling directions forever living side by side. We published our results in 2007 in the *Journal of Evolutionary Biology* as the first example of sexual selection and genital shape affecting the blueprint (clockwise or counterclockwise) of the entire body in an animal. And I present it here as the final example that genitalia hold the mirror up to nature.

What have we learned from this tour through the bizarre world of hermaphrodites? First, that their genitals are every bit as weird as those of "regular" animals. In fact, they are often even weirder and more exaggerated. Contrary to what Darwin thought, sexual selection holds sway in their evolution as well. Although every amorous slug or snail is wired the same way as the object of its desire, this does not lead to mundane, matter-of-fact, "you want it and I want it" sex. Quite the contrary: since each prefers the male role, they have evolved a whole battery of tricks to bring out, as it were, the feminine side of their partner. Love darts, sperm transferred at the ends of exaggeratedly long penises, sperm nibbling, and ridiculously large spermatophores are all strategies that have evolved to persuade the other to be the one that accepts more sperm. The outcome of this struggle is an amplified and more manipulative version of what we have already seen in gonochorists. The pinnacle, if you will, of what genital evolution can do to a species, right up to and including turning your whole body back to front.

Blushing is the most peculiar and most human of all expressions.

—Charles Darwin, *The Expression of the Emotions in Man and Animals*

Afterplay

When Jonathan Waage published his groundbreaking 1979 *Science* article in which he described the damselfly penis gizmo, he received some laudatory media coverage. But one headline—he thinks it was in the *National Enquirer*—blasted, "University Egghead Wastes Taxpayers' Money Studying Dragonfly Sex." Thirty-five years of genitalia research since Waage's landmark piece have not lessened the aim that certain media will take at its practitioners. In March 2013, the news network CNS derided the study on duck genitalia by Patricia Brennan and colleagues (Chapter 6) under a similar heading, leading to a two-week media mini-cyclone in the United States (outrage on Fox News; defense from *National Geographic* and *Slate*) that was quickly dubbed "Duckpenisgate."

It is easy to see why genitalia research is such an easy target (a sitting duck, almost) for those who wish to ridicule basic scientific research. Studying the finer details of the private parts of some inconsequential animal can effortlessly be cast in terms of frivolity beyond words. The reasoning would be a very effective double whammy of the following form: "Not only does nobody in the world, besides perhaps Patricia Brennan and her deviant colleagues at Yale, care about the exact shape of the genitals of the ruddy duck, but it also serves no practical purpose whatsoever—so public money should not be wasted on it!" No wonder 90 percent of the people polled by Fox News agreed with the latter conclusion.

Some arguments can in fact be brought to bear against the notion that genitalia research has no application. At several places in this book, we have seen cases of human and livestock fertility problems that can be understood—and sometimes solved—only with a good command of the evolution of the genitalia involved. In Chapter 4, for example, we saw that artificial insemination in domestic animals is improved if the shape of the syringe used resembles that of the penis of the respective species; Eberhard wrote a paper in the journal *Medical Hypotheses* arguing that the same might be true for humans.

The fact that that paper has been completely ignored by the medical literature (it has been cited exactly zero times since it was published in 1991) is symptomatic of the medical tradition of viewing human reproduction as a mundane bodily function, where both sexes are cooperating "for the good of the species," and fertility problems are seen simply as defects in that function. It should now be clear that reproduction in any species is the outcome of a never-ending evolutionary tango, a dance marathon that conserves elements of both battle and ballet. And, as we have seen, gynecological and urological problems in humans, such as preeclampsia (Chapter 7), spontaneous abortion (Chapter 4), seminal allergy (Chapter 7), and pregnancy in noncommunicating uterine horns (Chapter 6), which physicians often tend to take for granted as unavoidable "errors," can all be understood, and possibly better addressed, if we know the forces involved in genital evolution.

Yet potential practical applications should never be given by genital researchers as the sole justification for their work. Like art, music, and sports, basic science exists to provide a form of entertainment for the rest of humankind. Evolutionary biologists do the hard, painstaking labor to find out and make sense of all the details required to tell true stories about the world we live in. Nature documentaries on TV, for example, are watched by hundreds of millions of people worldwide. They marvel at and are entertained by what they learn. What is often forgotten is that it is not the presenters who, panting their way through steamy tropical jungles and along precipitous cliffs, have discovered all the biological facts that they are dishing out to the viewer. Every minute in such a documentary is based on years of work by some anonymous biologist doing fundamental scientific research sometime, somewhere.

By feeding (while at the same time being fed by) the deep-rooted human fascination with sex, genital researchers should have no problems getting the rest of the world to take them seriously. But therein also lies their weakness. For that deep-rooted fascination with sex is a curious beast. Since our genitals take up a strategic position in multiple strands of human emotion, we have always had a tumultuous relationship with our private parts. They pique our interest in a very urgent way, but they also are the seat of power and vulnerability in both men and women. It may well be that the curious mix of modesty and lewdness that results is an evolved aspect of human behavior that, in some ways, we share with our closest relatives. (Apes seem to know the special meaning of genitals, too. I have always been struck by Frans de Waal's observation that when two up-and-coming chimpanzee males in a colony staged a coup against the incumbent alpha male, they did not throttle him or bash in his head, but instead chose emasculation.)

Not surprising, then, that by drawing our attention to the origins, workings, and evolution of our and other species' reproductive organs, genital researchers get both the gain and the pain from this double-edged sword. Communicating the science involved is easy, because people instantly prick up their ears, but curiosity can quickly turn to embarrassment. When in 2013 *Scientific American* reported, via its Facebook page, on Brian Mautz's study on attractiveness of penis size in men (see Chapter 3), Facebook removed the (computer-generated) images for being too explicit.

At the same time, genitalia research itself is not free from entrenched notions about genitals and sex. Throughout this book, I have pointed to an ongoing debate between two camps of genital researchers. One camp, headed by Bill Eberhard, emphasizes the importance of cryptic female choice (CFC) for the evolution of genitalia. Other researchers, like Göran Arnqvist, think that sexually antagonistic coevolution (SAC) has been the driving force for the biodiversity of genitals. The differences are subtle. If CFC holds sway, then genitalia evolve mostly because females pick the best males from among the available suitors. Even if these males play vicious tricks to manipulate the female, in the end it is the female who "selectively cooperates," as Eberhard puts it, to provide her offspring with the best genes she can find. The SAC camp, on the other hand, sees genital evolution as going off down a vortex of male-female conflict over who has the final say in fertilization—with both parties suffering.

It is hard to decide which camp is right. The only way to obtain hard evidence would be by comparing, under natural circumstances, the exact numbers of children and grandchildren and great-grandchildren—in other words, the total reproductive success—of females that mate with different types of males. If these numbers go down when females mate with many males, then that would be evidence for SAC; if they go up, CFC would have the best cards. But nobody "has ever come even close to producing reasonable data on this score," according to Eberhard.

Meanwhile, both camps accuse each other of falling victim to old, nonscientific, patriarchal stereotypes about sex roles. CFC adept Eberhard speaks of "cultural blinders" that unconsciously persuade SAC folk that sexual interactions are still more about males being active and females being passive. The SAC camp, on the other hand, points out that CFC people are too much penis-focused and often ignore the gynecology of their favorite species.

The charge of penis staring and vagina neglect should certainly be taken to heart by the whole field of genital science. Partly for practical reasons (in insects, for example, the penis is usually made from chitin and retains its shape even in dried and pinned specimens, whereas the soft and pliable female genitals shrivel), penises have been favored and vaginas shunned. But there must also be a cultural bias, as we have seen from the struggle to get the human clitoris properly described, compared with the penis, which never had any trouble claiming a chapter for itself in the textbooks. As feminist evolutionary biologist Patricia Adair Gowaty has written, female genitalia, throughout biology, have suffered from being viewed as "more common than elaborate, more utilitarian than bizarre." Yet, more often than not, whenever people have overcome the practical and cultural obstacles to take a good look at female genitalia (think of the ducks and rove beetles of Chapter 6), they appear to be just as peculiar and multiform as the male ones. In fact, in many types of animals (spiders, clown and featherwing beetles, sand flies, and moths, to name but a few) the female genitals are used for species identification just as the male ones are.

Despite such shortcomings and biases, studying the evolution of genitalia has provided us with deep insights and spectacular panoramas on the history of life. When Bill Eberhard, as a young student, began flipping through arachnological monographs, what ignited his interest was that no other body part has been diversified by evolution as much as the organs directly involved in reproduction. The fact that genitalia are the acme of biodiversity in many kinds of organisms seems almost like a law of nature. But let us not forget the tenuous and contingent routes that have led to it.

Somewhere in the deep folds of time, sex arose as a means to outrun fast-evolving viruses or help fix errors in the genetic code. Then a bacterium crawled into a primordial cell. To counter combat among cohabiting bacteria, separate male and female sexes, producing different sex cells (eggs and sperm), appeared. Rather than scattering these sex cells randomly in the currents, some organisms began packaging them in spermatophores and evolved genital organs to hand these to one another, setting the stage for sexual selection to wave its magic wand of cryptic female choice, Fisherian selection, and sexual conflict, and giving us the bewildering array of form and function that we see today.

But there is more. The proteins that form the mating plugs of Chapter 7 are possibly the same ones initially used to build the casings of the spermatophores in the ancestors of these animals, perhaps even before genitalia evolved. Likewise, the cell-penetrating qualities of spermatozoa came in handy in those animals where traumatic insemination evolved, when it became necessary for sperm cells to penetrate more than just the outer layer of an egg cell. The fact that hermaphrodites produce female hormones as well as male ones preadapts them for the hormonal manipulation of their mates, as Joris Koene pointed out to us in Chapter 8. And, finally, there may be a finer line between the sperm dumping of Chapter 3 and the sperm scooping of Chapter 6. A male, during copulation, may be able to trigger the reflex that makes a female dump a previous male's sperm, blurring the distinction between sperm dumping and scooping.

Such contingencies and predestinations form a loose straitjacket for genital evolution. Loose because, as we saw in Chapter 4, within certain confines genital evolution is such a dynamic process that it can go off in all kinds of unpredictable directions, leading to great multiformity. At the same time, environment plays a much bigger role than just curbing the size of the genitals of mosquito fish and *Tidarren* spiders. I am writing this section while on vacation in Crete. Unlike most vacationers, I have spent some time looking at copulating beetles on roadside flowers. While the copulas of some species seem oblivious to the outside world and stay firmly locked together as I pick them up and put them under my hand lens, others separate and fly off instantly when disturbed. Perhaps the former kind are locked together by spines on the penis, and perhaps these could evolve because these beetles are toxic or for some other reason have fewer predators and hence are less vulnerable when caught in flagrante delicto. I would not be surprised if such evolutionary communications exist between a species' genitals and the niche it lives in.

Conversely, genital evolution may also be the driving force for the origin of new species. If it is true that male genitalia are constantly evolving to adapt to female genitalia and vice versa, then this dual adaptation could cause such splitting, so-called speciation events. In the substance-injecting seed beetles that we met in Chapter 7, for example, different species have different degrees of penis spininess and the vaginas are correspondingly thicker skinned, such that females of species in which the males have particularly spiny penises have more heavily armor-plated vaginas. Such mutual adaptations do not just happen in seed beetles that already belong to separate species; the same process also takes place in separate populations of the same species—for example, in the seed beetle *Callosobruchus maculatus.* Males of an Indian *C. maculatus* that mate with an African female are not very successful in the sperm competition with males from the female's native population, and vice versa. Such mutual male-female adaptation taking place in each population separately could give rise to barriers to effective crossbreeding between populations, which can sometimes cause these populations to evolve into separate species.

But that's seed beetles. In the Preliminaries, I pointed out that there are striking differences between the genitalia of ourselves and our closest relative, the chimpanzee. And while I would not imply that the evolution of genitalia is the wedge driven between us and our primate brethren, it certainly has played its part in paving the evolutionary paths that these species have traveled over the past few millions of years. Throughout this book, we have seen tantalizing glimpses of that: penises that may have lost their spininess but have evolved shapes that help displace sperm, and female orgasms that select or reject semen from males, which, in turn, may be subtly modifying female physiology or even hampering other males' semen.

What I haven't discussed—genital researchers haven't even begun scratching the surface of it—is the possibility of homosexuality also playing its part in genital evolution. In mammals, including humans, this is likely to be relevant. For example, the glans of the clitoris in bonobos is larger and more forward pointing, an anatomical feature that some primatologists have linked with the fact that homosexual copulations among females (as well as among males) play an important social role in those apes. We have also come across the theory (in Chapter 6) that homosexuality in male bedbugs helps a male to sire offspring by proxy, via the male mate who then goes on to mate with females—if you can still follow. With homosexuality rife in the animal kingdom, it is likely that other ways exist in which same-sex mating may have an impact on genital evolution—none of which has yet been investigated by biologists.

There is a vast ocean of sexual function beyond the quiet backwater that we humans find ourselves in, and much of what exists in other animals is extremely foreign to us: we would not be very happy if we were to live the sex lives of seed beetles, sea slugs, or cellar spiders. But what I hope to have shown is that humans *do* fit in. The ways in which the shapes and functions of our genitals have been molded by evolution merge seamlessly with the rest of the continuum, making us an integral part of nature. With all the animals we share this world with, we are descendants of an unbroken line of ancestors who, like that seventeenth-century couple inspecting a sperm whale penis, all agreed on the irresistible lure of sex.

Acknowledgments

I wrote this book partly at home, partly while traveling in different parts of the world. At home, I want to thank my partner, Rachel Esner, and my children, Fenna and Jan, for accommodating me and my writing-enhanced pensive moods at the kitchen table in Leiden, Amsterdam, Paris, and Kota Kinabalu—and Rachel for allowing me to muse on genitalia even more than I normally do. Abroad, unbeknownst to them, the staff of the following places helped nurse this book to life. In London, the Dana Centre, the Imperial College canteens, the Victoria and Albert Museum restaurant, and Mrs. Sue Crayson's bed-and-breakfast. In the Azores, the Lince Hotel in Ponta Delgada. In Crete, the Marin Hotel and Morosini café in Iráklion (overlooking the city's bustling lek), and in Doulianá, Filippa's House. The whole of Chapter 6 was written in Ben Bankson's apartment in Brooklyn, New York, and much of Chapters 1 and 2 in the apartment rented from Annette Ciré in Berlin.

Then there are the many scientists (and the occasional artist and art historian), some of them friends, others colleagues, yet others fully unacquainted to me, who all freely donated smaller and larger bits of their time, offered advice and information, and in many cases hospitably allowed me into their labs, offices, and homes. In alphabetical order, those were: Gerd Alberti, Göran Arnqvist, Lisa Becking, Tim Birkhead, Oskar Brattström, Maria Fernanda Cardoso, Satoshi Chiba, Bill Eberhard, Coby Feijen, Hans Feijen, Claudia Gack, John Grahame, Martin Haase, Peter van Helsdingen, Kasper Hendriks, Rolf Hoekstra, Michael Jennions, Kazuki Kimura, Joris Koene, Hanna Kokko, Bram Kuijper, Michael Lang, Martine Maan, Patricia Mainardi, Brian Mautz, Thibaut de Meulemeester, Peter Michalik, Jeremy Miller, Kees Moeliker, Nobuaki Nagata, Virginie Orgogozo, Antonella di Palma, Rich Palmer, Michel Perreau, Michal Polak, Andrew

Pomiankowski, Richard Preece, "Theo" Michael Schmitt, Angela Schmitz-Ornés, Stephen Sutton, Gabriele Uhl, Jonathan Waage, and Paul Watson.

Several friends have given advice and intellectual as well as mental support, especially, and in many, many ways, my partner Rachel Esner, but also Frank van Rooij (both proofread large parts of the draft), Jeroen Roelfsema, Janine Kourakos, Lydian ter Brugge, and Abigail Solomon Godeau. I would also like to acknowledge my colleagues at Naturalis Biodiversity Center, especially Kris de Greef and all the members of the focus group Character Evolution, my genitalia students Flor Rhebergen, Ruben Vijverberg, Thijmen Breeschoten, Paulien de Jong, Tamara Hoogenboom, Rick van Beek, and Melanie Meijer zu Schlochtern, and also my directors Jan van Tol, Koos Biesmeijer, Erik Smets, and Edwin van Huis, whose flexibility has allowed me to divert part of my time to this book project.

Finally, Peter Tallack and Louisa Pritchard at the Science Factory and Melanie Tortoroli and Wendy Wolf at Viking, who have made it all possible, with the inestimable assistance of designers Francesca Belanger and John Patrick Thomas, senior production editor Bruce Giffords, and illustrator Jaap Vermeulen.

Notes

Preliminaries

1 Nowadays, the **Netherlands Natural History Museum** is housed in a modern building at the Bio Science Park of Leiden—and has been renamed Naturalis Biodiversity Center (www.naturalis.nl). The collection of antlers and the whale painting (depicting a beached sperm whale on the beach of Brouwershaven, 1606) are also there.

1 *The Sex Life of Wild Animals* is Burns (1953).

1 The classroom wall poster *Penises of the Animal Kingdom* was produced by Jim Knowlton (who received an Ig Nobel Prize for it in 1992).

1 The online series *Green Porno* can be viewed via www.sundancechannel.com/greenporno.

2 **Darwin's (1871) denial** of primary sexual characteristics as being under sexual selection is outlined on pp. 253–54 of his Chapter 8 ("Principles of Sexual Selection").

2–6 **Waage (1979) and Eberhard (1985)** are listed in the bibliography. The conditions under which Waage began **his damselfly genital studies** are from an online interview I conducted with him on March 14, 2013.

3–4 The events surrounding the conception of **Eberhard's book** are from an e-mail correspondence with Bill Eberhard in late April 2013.

4 The **comment by Stephen Hubbell** is from my book *The Loom of Life* (Schilthuizen, 2009).

5 **The female genitalia of the chimpanzee** is described in Dahl (1985).

5 **The male genitalia of the chimpanzee** is described in Prasad (1970).

6 A paper in *Nature* (Barron and Brown, 2012) attacks the tendency for **sensationalism and misreporting of sexual selection research** by the science media.

Chapter 1: Define Your Terms!

9 Current views on **bacterial sex** are described in Prasad Narra and Ochman (2006).

10 **Mass spawning in coral** is described in, for example, Harrison et al. (1984).

10 Descriptions of **sperm transfer without direct contact between males and females** is described for pseudoscorpions, springtails (Collembola), and salamanders in Weygoldt (1969), Schaller (1971), and Houck and Verrell (2010), respectively.

10 A good overview of the different **sperm transfer methods among animals** is Clark (1981).

10 The theory that **spermatophore deposition predated the evolution of genitalia** is to be found in several chapters in Highnam (1964).

10–11 More information on **asexual reproduction** in plants, parasitic wasps, stick insects, beetles, turkeys, lizards, and bdelloid rotifers can be found in Calzada et al. (1996), Godfray (1994), Bullini and Nascetti (1990), Dybas (1978), Olsen (1966), Radtkey et al. (1995), and Fontaneto et al. (2007), respectively.

11–12 The various theories for the **origin of sexual reproduction** are explained in a very accessible way by Ridley (1993).

13–14 Background on the **evolution of the sexes** is in Hoekstra (1987), Randerson and Hurst (2001), and Schilthuizen (2004). This and the previous section were checked by Rolf Hoekstra.

14–15 A representative medical definition of **primary vs. secondary sexual characteristics** may be found in Haeberle (1983).

15 Darwin (1871, Part II: 253–55) describes Hunter's (1780) **primary and secondary characteristics and problems in defining them in animals.** A modern treatment of the same is in Ghiselin (2010).

15 **The genitalia of *Cycloneda sanguinea*** are described in Eberhard (1985: 158) and Araújo-Siqueira and Almeida (2006).

16 The **blue testicles** of male l'Hoest monkeys and mouse opossums are described in Prum and Torres (2004).

16–17 More background on the **discovery of sexual selection** can be found in Chapter 4 of my book *Frogs, Flies, and Dandelions* (Schilthuizen, 2000).

17 Ghiselin's **critique of the use of the term "character"** is in, for example, Ghiselin (1984).

18 Eberhard's discussion of the **definition of genitalia** is in Eberhard (1985: 2).

18–19 A good discussion on the **evolution of internal fertilization** as a result of life on land is Clark (1981).

19 Van Leeuwenhoek (1678) was the first to describe (in Latin, because of the sensitive nature of the subject) the **death of dog sperm cells in freshwater.**

19 A detailed figure showing the life span of various **fish species' sperm cells** in distilled water of a range of temperatures is in Lindroth (1947).

20–21 The **medical cases of "infection" with squid sperm** are described and interpreted zoologically by Marian et al. (2012) and also by Danna Staal in a videotaped performance on Nerd Night San Francisco (June 20, 2012). The headline quoted was from the June 15, 2012, online issue of the UK's *Daily Mail,* based on a paper by Park et al. (2012).

21 Tinbergen (1939) describes the **mating behavior** in *Sepia* cuttlefish.

23 The information on **hectocotylus autotomy** in *Argonauta* is from Laptikhovsky and Salman (2003) and Sukhsangchan and Nabhitabhat (2007).

23 The **argonaut shell,** which I briefly allude to, is a source of much speculation, since it is so similar to an ammonite shell and yet is made of a different material (calcite vs. aragonite) and not by the animal's mantle, but by its tentacles. For this reason Naef (1921) has suggested that, when ammonites were not yet extinct, the ancestors of argonauts began using discarded ammonite shells for laying their eggs in, and over time evolved the ability to repair them and, eventually, to handcraft them themselves. This fanciful suggestion is not taken seriously by many, though. Another fascinating aspect of argonaut life is that they sometimes seem to live in symbiosis with jellyfish (the females) and salps (the males). A good video with basic information on argonauts is at museumvictoria.com.au/about/mv-news/2010/argonaut-buoyancy/.

23 Cuvier's description of *Hectocotyle octopodis* is in Cuvier (1829).

24 **Sedgwick's paper** read to the Royal Society is Sedgwick (1885). **Group hunting in social onychophorans** was described by Reinhard and Rowell (2005).

25–26 The **strange new head structures** were first reported by Tait and Briscoe (1990). Tait and Norman (2001) described the mating behavior in *Florelliceps stutchburyae.* The discovery that onychophorans with head structures also maintain organs for fertilization through the skin is in Walker et al. (2006).

Chapter 2: Darwin's Peep Show

The first seven hundred words of this chapter were published, in Dutch and in altered form, as Schilthuizen (2013).

28–29 My **visit to the Jardin des Plantes** and the Muséum National d'Histoire Naturelle in Paris, in search of René Jeannel's statue, took place on July 16, 2012.

29 Information on **Jeannel's biography** I took from Motas (1966), Negrea (2007), and d'Aguilar (2007), and from the page on René Jeannel at fr.wikipedia.org/wiki/René_Jeannel.

29 The information on **cave art** discovered by Jeannel is mentioned in Lawson (2012).

29–30 **Jeannel's doctoral thesis** is Jeannel (1911). Leptodirini, a tribe in the small fungus beetle family Leiodidae, were in Jeannel's days called Bathysciinae and considered to be a subfamily of the carrion beetles, Silphidae.

31 Jeannel's **monograph of the "aedeagus"** is Jeannel (1955).

31–32 Examples of **bumblebee taxonomy** are Williams (1991) and Richards (1927), although the species specificity of the bumblebee penis was already noted by V. Audoin in 1821 (Jeannel, 1955).

31–32 More information on **bumblebee genitalia** can be found at www.nhm.ac.uk/research-curation/research/projects/bombus/genitalia.html.

31–32 More **information on bumblebees** in general can be found at www.bumblebee.org and bumblebeeconservation.org. Although bumblebees predominate in the Northern Hemisphere, some species occur in South America and the mountainous parts of Southeast Asia.

32–34 Some background on **elephant shrews** and their classification can be found in Springer et al. (1997) and Dawkins (2004: 224); the study on male genital shapes in these animals is Woodall (1995), whereas some background on the female reproductive system is in Tripp (1971).

34 In Chapter 4 of my book *Frogs, Flies, and Dandelions* (Schilthuizen, 2000), I place **genital diversity in the context of speciation** (see also Afterplay in the present book).

34 The **entomologist who impressed me** at a Netherlands Entomological Society meeting (in the winter of 1981–1982) was Mr. Diakonoff.

34–35 The paragraphs on **Maria Cardoso** were based on a documentary by the Australian Broadcasting Corporation, online at www.abc.net.au/arts/artists/maria-fernanda-cardoso-the-museum-of-copulatory-organs/default.htm, the artist's own Web site, www.mariafernandacardoso.com, and e-mail exchanges with her throughout the writing of this book.

36 **Gosse's quote** is from Gosse (1883). I must admit that I, as many people before me, have quoted Gosse deliberately out of context. Gosse actually goes on to state that although the view of genitalia as locks and keys is often expressed, "I should like to see these axioms demonstrated."

36 Further background on the **lock-and-key hypothesis** can be found in, for example, Eberhard (1985) and Shapiro and Porter (1989).

36 In this section, I combine the **"mechanical" and the "sensory" versions of the lock-and-key hypothesis.** The latter is worded in, for example, Jeannel (1941, 1955) and De Wilde (1964).

36 The paper on **carrion beetles** in a Czech forest is by Kočárek (2002), and illustrations of the penises of *Catops* species may be found in, for example, Jeannel (1936).

37 Fooden (1967) described the **macaque genitalia.**

37 **Lock-and-key theories were questioned** for bumblebees by, for example, Richards (1927) and Boulangé (1924).

37 The entomologist who showed me his **cross-species copulas** was the late Jan Lucas.

39 The mating experiments on **chiral *Ciulfina*** are described in Holwell and Herberstein (2010).

39–40 The **biogeographic evidence** against the lock-and-key hypothesis is in Eberhard (1985), but also Shapiro and Porter (1989), Eberhard (1996), Arnqvist (1997), and Eberhard and Huber (2010).

40–41 The example of **lock-and-key-like patterns in** *Drosophila* is from Kamimura and Mitsumoto (2012).

Chapter 3: An Internal Courtship Device

42–43 The **field guides** mentioned are Jeyarajasingam and Pearson (1999), Kleukers et al. (1997), and Freude et al. (1971).

44 **Eberhard's quote** is from Eberhard (1985: 15).

44 The **Darwin caricature** is by Albert Way (1805–1874) and is held by the Cambridge University Library. (As far as we know Darwin never rode any giant beetle—only giant tortoises.) The account of his beetle-collecting days is from Barlow (1958: 62–63).

44 **Darwin's barnacle treatises** are Darwin (1851, 1854).

44 In Darwin's 1854 monograph, he describes the **penis of the burrowing barnacle** (according to Birkhead, 2009) as "wonderfully developed . . . when fully extended, it must equal between eight and nine times the entire length of the animal . . . coiled up, like a great worm."

44 The list of **longest penises in the animal kingdom** is in Neufeld and Palmer (2008).

44 More details on **barnacle penis morphology** are in Hoch (2008) and Neufeld and Palmer (2008), while a video of mating barnacles can be found at vimeo.com/7461478.

45–46 The **influence of Henrietta Darwin** on her father's work is described in, for example, Birkhead (2007, 2008, 2009), while the anecdote on her stinkhorn campaign comes from Raverat (1952) (*fide* Birkhead, 2007). I also used notes made during Birkhead's Tinbergen Lecture at Leiden University on May 11, 2012. Her diary, as well as a commentary on it, is published at www.darwinproject.ac.uk/hed-diary-1871. Henrietta's denial of the "Lady Hope Story" of Darwin's deathbed change of heart is in Litchfield (1922).

47–48 **Bateman (1948)** became a key paper in the development of sexual selection theory after it was "rediscovered" by Trivers (1972). However, Bateman (1948) made errors in his experimental design and in the analysis of his results, which were highlighted by Gowaty et al. (2012), who also repeated his experiments and found no evidence for sexual selection. Other criticisms of Bateman's principle have been summarized by Judson (2002). The section on Bateman's principle was read and commented upon by Hanna Kokko, who shared with me two unpublished (partial) book chapters by her on the subject, and also pointed me to the paper by Gerlach et al. (2012), which explains why positive Bateman gradients for females might be statistical artifacts.

48–50 A popular account of the **history of sexual selection**, especially the "female choice" version of it, is in Schilthuizen (2000). In that book, I adhere to the conceptual difference between "good genes" and "Fisherian" sexual selection—see also Ridley (1993). However, as, for example, Kokko et al. (2003) have shown, there is no basic difference between the two.

50–51 The work on **sexual selection in poison frogs** is in Maan and Cummings (2008, 2009, 2012). For more information on this system, see Myers and Daly (1976) and Summers (2004).

51 The examples of **female choice in peacocks and barn swallows** are from Petrie (1994) and Møller (1990), respectively.

51 The original discovery of the **effect of leg bands** on female choice is in Burley et al. (1982).

51–52 Burley's work on the **effect of head crests** is in Burley and Symanski (1998), whereas the information on a change in behavior of males made artificially attractive is in Burley (1988).

52–53 **Sensory drive** is a complex process with several components that have been separately studied and named. Endler and Basolo (1998) paint a coherent picture of all this.

53–54 **Visual penis display** in mosquito fish, primates, and lizards is described in, respectively, Kahn et al. (2010), Wickler (1966), and Bohme (1983), whereas Mautz et al. (2013) is the reference for the study on human penis length as a factor in male visual attractiveness. The squirrel monkey example comes from Ploog and MacLean (1963).

54 The cartoons of **Rube Goldberg** can be found at rubegoldberg.com.

54–56 The paper on **stridulating genitals** in crane flies is Eberhard and Gelhaus (2009), whereas papers on genital stridulation in moths are, for example, Heller and Krahe (1994) and Conner (1999).

55–57 The function of **parameres** in general is described in Eberhard (1985) and more specifically in beetles in Düngelhoef and Schmitt (2010). Please note: the sentence placed in quotation marks ("more to the left—yes, that's it") is not intended to be quoting Dr. Düngelhoef. Parameres in the ladybird beetle *Cycloneda* are described in, for example, Araújo-Siqueira and Almeida (2006) and a video of a mating pair (with paramere action!) is at www.youtube.com/watch?v=VplJpd lUmsw.

57 The **penile flagella** in insects are described in Eberhard (1985) and Rodriguez et al. (2004), while the urethral processes in hoofed mammals are in Prasad (1970). The report that in cow copulation this process flips forward is from Eberhard (1985: 11).

57 The cultural anthropology of the **Bornean *palang*** is described with relish in Harrisson (1959); the Sarawak Museum in Malaysia holds several famed examples.

57 The **quotation on wasp genitalia** is from West-Eberhard (1984).

57–58 The proportions of species in which **rhythmic genital thrusting** occurs, including the detail on the bush baby, as well as the observations on "cryptic thrusting," are taken from Eberhard (1996).

58 The **phalloid organ** of the red-billed buffalo weaver and its use in copulation is described in Winterbottom et al. (1999, 2001) and in the Tinbergen Lecture by Tim Birkhead at Leiden University on May 11, 2012.

58 The **sense organs** in the damselfly vagina, cockroach female genitals, human clitoris, and human vagina are derived from Córdoba-Aguilar (2005), Eberhard (2010a), Cold and McGrath (1999), and Pauls et al. (2006), respectively.

58–59 An example of the **heritability of genital shape** is Sasabe et al. (2007).

59–61 **The two back-to-back papers** in *Animal Behaviour* are Baker and Bellis (1993a, b), and the details in this section are from these two papers, as well as from a conversation I had with Robin Baker during the Sixth Congress of the European Society for Evolutionary Biology, Arnhem, August 24–28, 1997.

59–60 **Baker's later work has been criticized** by, among others, Tim Birkhead, first mildly (Birkhead, 1995), later more vehemently (Birkhead et al., 1997).

61 The quotation on **Baker and Bellis's popularity** during the Princeton congress is from Birkhead (1995).

61–62 The **female genitalia of the spider** *Silhouettella loricatula* and how they might function in the species' habit of sperm dumping are described in Burger et al. (2006) and Burger (2007).

62–64 **Sperm dumping** in cellar spiders is described in Peretti and Eberhard (2009), in Grevy's zebra in Ginsberg and Huck (1989) and Ginsberg and Rubenstein (1990), and in chickens in Dean et al. (2011). The description of the copulation and sperm dumping in the nematode worm *Caenorhabditis elegans* is in Barker (1994). The thrusting movements of nematode spicules during copulation, as well as further details about mating, are described in Barr and Garcia (2006).

Chapter 4: Fifty Ways to Peeve Your Lover

65 The **quotes from Eberhard's book** are all from Eberhard (1996).

66 Some of **Donald Dewsbury's papers** that I consulted are Dewsbury (1971, 1974) and Langtimm and Dewsbury (1991).

67 Another early author to describe a **"dry" phase in spider copulation** was Kullmann (1964).

67 **Dry sex** in millipedes is described in, for example, Tanabe and Sota (2008). The examples of other animals are taken from Eberhard (1985).

67–68 I interviewed **Peter van Helsdingen** in Leiden on October 25, 2012. His paper is Van Helsdingen (1965).

67–71 The **female reproductive morphology** for several species of sheet web spiders is described in, for example, Hormiga and Scharff (2005) and for the golden hamster in Yanagimachi and Chang (1963).

68–70 **The work on** ***Neriene litigiosa*** is described in, for example, Watson (1991, 1995) and Watson and Lighton (1994), as well as in Eberhard (1996). Further details are based on an e-mail from Paul Watson of October 28, 2012. In reality, sierra dome spiders do not very often opt out by departing after the dry phase of copulation, instead relying more on sperm dumping in most cases.

73 For the positive effects of proper **female stimulation in artificial insemination**, see Evans and McKenna (1986) and Eberhard (1991). Many more eyewitness accounts on artificial insemination in pigs are in Roach (2008).

73 The work on the **spotted cucumber beetle** is in Tallamy et al. (2001).

73–76 I took most of the information on the anatomy of the **human clitoris** and the history of its study from O'Connell et al. (1998, 2005) and Foldes and Buisson (2009). Particular details on Reinier de Graaf were taken from Rozendaal (2006), while the *New Scientist* article is Williamson (1998).

74 The dissections of the clitoris by Kobelt are described in Kobelt (1844).

75–76 My brief paragraph on the changing views on **human female orgasm** is largely based on Gould (1991) and Symons (1979). Freud's ideas on female orgasm are in the third of his *Three Essays on the Theory of Sexuality* (Freud, 1905). The quote from *The Naked Ape* is from Morris (1967).

76 The development of **marsupial genitals** is taken from Butler et al. (1999).

76 Information on **clitoris shape in several mammals** was taken from Porto et al. (2010), Place and Glickman (2004), Rubenstein et al. (2003), and Bassett (1961).

77 The **quote by Tim Birkhead** is from his Tinbergen Lecture at Leiden University on May 11, 2012.

77 For the physiology of **vaginal and clitoral orgasms**, see Masters and Johnson (1966) and Gruenwald et al. (2007).

77 The release of **oxytocin during orgasm** is from Blaicher et al. (1999).

77 Information on brain regions activated during orgasm is from Georgiadis (2011).

77 The **urethrogenital reflex** in female rats is described in, for example, Marson et al. (2003) and Giraldi et al. (2004).

77–78 The reference for the work on **pressure changes in the uteri of cows** during copulation is VanDemark and Hays (1952).

78 For **female orgasm in primates**, see Burton (1971), Hanby and Brown (1974), Symons (1979), Puts et al. (2012a), and Troisi and Carosi (1998).

78–79 Other **upsuck experiments** on mammals, even nineteenth-century ones, can be found in Roach (2008). The references used for the bit about the upsuck hypothesis are Fox et al. (1970) and Baker and Bellis (1993b).

80 The studies that used questionnaires to look for the effects of **male attractiveness and penis length** on orgasm frequency are Puts et al. (2012b) and Costa et al. (2012).

80–81 The **by-product hypothesis** of the female orgasm is first coined in Symons (1979). The 0.8-second interval in both male and female orgasmic spasms is from Masters and Johnson (1966). Gould's "clitoral ripples" essay is included in Gould (1991), and other recent publications in favor of the by-product hypothesis are Lloyd (2005) and Wallen and Lloyd (2008). See also Lloyd's 2012 lecture at the Istituto Veneto, available at www.youtube.com/watch?v=m6GMeeOFUsE. Rebuttals of the by-product theory are in, for instance, Judson (2005), Hosken (2008), Lynch (2008), Zietsch and Santtila (2011), and Puts et al. (2012a).

81 The **role of oxytocin in sperm transport** is in Wildt et al. (1998).

81 I got the number of **titles on the biology of the female orgasm** in the years 2008–2012 by searching for "female orgasm" AND "biology" in Google Scholar.

82 One story of a **woman impregnating herself** with her ex-husband's sperm appeared in the UK's *Daily Mail* on September 24, 2012.

82–83 Lists of **sperm longevity** in various kinds of vertebrate animals are to be found in Birkhead and Møller (1993) and Holt and Lloyd (2010), and for social insects I used Boomsma et al. (2005).

83 Eberhard (1996) lists a number of different **sperm storage organs** in female animals, including snakes, and Pitnick et al. (1999) give background information for flies and other insects. For snails, I used Baur (2007, 2010) and Evanno and Madec (2007), and for turtles, Gist and Jones (1989).

84–87 For the **story about the yellow dung fly**, I primarily used Parker (2001), from which also Parker's quotes are taken, Ward (1993),

Hellriegel and Bernasconi (2000), Jann et al. (2000), Sbilordo et al. (2009), and Bussière et al. (2010).

87 The **Bruce effect** was described by Bruce (1959, 1960). The examples from the Bruce effect in various vole species I took from Eberhard (1996: 164). The study on the Bruce effect in wild geladas is Roberts et al. (2012), while I also used Yong (2012).

88 The information on **natural abortion rates in humans** is from Forbes (1997), Reeder (2003), and Wasser and Isenberg (1986), while the correlation between abortion and immune system similarity is in Knapp et al. (1996) and Ober et al. (1998) for macaques and humans, respectively. The reference to antiabortionists refers to the remarks by U.S. Senate candidate Todd Akin that "the female body has ways to try to shut that whole thing down."

Chapter 5: A Fickle Sculptor

90 I took **general ideas on sexual selection** from Kuijper et al. (2012).

90 The **opening sentence quoted** ("Let *t* be a male trait used by females in mate choice and *p* be the strength of female preference") is from Iwasa and Pomiankowski (1995).

90–91 I took some general information on **stalk-eyed flies** from Cotton et al. (2010) and from interviews with Hans and Cobi Feijen during the Kinabalu/Crocker Range Expedition, September 2012.

91 The **field trip with A. Pomiankowski and S. Sutton** took place on April 17, 2006.

93–94 The **rare-male effect** is described in, for example, Partridge (1988) and Kokko et al. (2007).

93–94 The **guppy experiments,** from Hughes et al. (1999), were more complex (involving also a third, more variable group of males) than I have described them here, but by focusing only on the M1 and M7 males I hope I have represented their essence. The part of this section that deals with guppies was kindly checked by Michael Jennions.

94 The **haiku** by Hanna Kokko is one that I took from her home page, biology.anu.edu.au/hosted_sites/kokko/Publ/index.html (and I apologize to her for not displaying it in proper three-line format).

94–95 Her **computer modeling of the rare-male effect** is in Kokko et al. (2007). I made use of a lecture by and an interview with her in Leiden, February 15, 2013, and of comments she made after proofreading this section.

95–96 The **rapid change in Galápagos finches** and other contemporary evolution is described in, for example, Weiner (1994).

96 The book on the **paleontology of sex** is Long (2012).

96 The **fossilized turtle copulation** is from Joyce et al. (2012), and the reproduction in extinct placoderms is in Long et al. (2008, 2009).

96 The **history of amber insect collections** is in Poinar (1992).

96–97 The **synchrotron X-ray tomographic technique**, as well as some results in imaging the genitalia of Baltic amber beetles, is in Perreau and Tafforeau (2011) and Perreau (2012). Michel Perreau read and checked the paragraphs on his work.

97–98 The use of **beetle fossils from peat** and other Pleistocene and Holocene deposits is advocated by Coope (1979, 2004) and Schafstall (2012), and its history was clarified to me by Richard Preece during a meeting at Cambridge, UK, April 17, 2013. Coope and Angus (1975) give illustrations showing the excellent preservation of genitalia in such fossils, and the quotation is from Coope (2004).

98–99 The original **punctuated equilibria** article is Eldredge and Gould (1972), whereas the terms "evolution by jerks" and "evolution by creeps" are mentioned in, for example, Jones (2008).

99 The **Darwin quote** is from *The Origin of Species* (Darwin, 1859).

99–100 The work on **3-D analysis of clasper shape evolution in damselflies** can be found in McPeek et al. (2008, 2009) and Shen et al. (2009).

100 The studies that show **jerky evolution in the genitals** of flies and millipedes are, respectively, Richmond et al. (2012) and Wojcieszek and Simmons (2011).

100–101 One of the citations that seem to have been the basis for the **"size doesn't matter" statement** is Masters and Johnson (1966: 91).

101 General background information on **allometry** is to be found in Klingenberg (1996).

101–3 Surveys of **negative allometry in the genitalia** of large numbers of animal species are provided by Eberhard et al. (1998) and Eberhard (2008), whereas the stag beetle example is in Tatsuta et al. (2001). Retief et al. (2013), however, suggest that positive allometry may be common in mammal genitalia.

102–3 The study of **human penis length and shoe size** is Shah and Christopher (2002), and the online survey is by Richard Edwards (www.sizesurvey.com).

104 **Allometry in guppy genitalia** is mentioned in Eberhard (2008) and Kelly et al. (2000).

104–5 The work on **mosquito fish** is in Langerhans et al. (2005) and in Kahn et al. (2010). The part of this section that deals with poeciliid fish was kindly checked by Michael Jennions.

105–7 **Background information on *Tidarren*** was taken from Knoflach and Van Harten (2000). The term "one-shot genital apparatus" is from Schneider and Michalik (2011). The evidence that the *Tidarren* testes wither before maturation are from Michalik et al. (2010). The experiments on mobility are in Ramos et al. (2004). What I wrote on *Tidarren* was read by Peter Michalik, except the sentence on the evolution of small males and large females in this genus, which was added afterward and was taken from Hormiga et al. (2000).

Chapter 6: Bateman Returns

109–12 The **functioning of the genitalia in damselflies** is from Waage (1979), Córdoba-Aguilar et al. (2003), and Cordero-Rivera and Córdoba-Aguilar (2010).

110–13 The **details on Waage's career** are from an e-mail correspondence I had with him on March 14, 2013, and from his CV available from his home page at Brown University.

113 **Waage's work on the spreadwing *L. vigilax*** is in Waage (1982).

113–14 The **information on *Calopteryx xanthostoma*** is from Siva-Jothy and Hooper (1996), and on *Calopteryx haemorrhoidalis* from Córdoba-Aguilar (1999, 2002).

114 **Sperm dumping in damselflies** is discussed in Eberhard (1996).

115 The **quote on swords and shields** is from Dawkins and Krebs (1979).

115–16 The **risks of anthropomorphizing sexual conflict** and choosing a human-gender-associated terminology are discussed in Karlsson Green and Madjidian (2011).

116–17 The **experiments with artificial human genitals** are in Gallup et al. (2003), whereas the results of their college student interviews are in Gallup et al. (2006).

116–17 More information on **the frequency of sexual encounters in women** is in Baker and Bellis (1995).

117 The data on **single- or multi-perpetrator rape** are, for example, in Vetten and Haffejee (2005).

117 The information on **chimpanzee promiscuity** is in Tutin (1979).

118 The Belgian-British **fertilization-by-proxy** article is Haubruge et al. (1999).

118–19 I obtained basic information on **shark genitalia and copulation** from Gilbert and Heath (1972), Fitzpatrick et al. (2012), and Eilperin (2012).

119 Whitney et al. (2004) published observations that they claim refute the **sperm-flushing hypothesis for sharks**, but I am not entirely convinced by their argumentation, which is why I have still decided to include sharks (with some caution) in the list of animals that do sperm displacement.

119–20 **Sperm removal in crickets and katydids** is in Von Helversen and Von Helversen (1991), Ono et al. (1989), and Eberhard (1996).

121–24 I thank Theo Schmitt for making me more fully aware of **the genitalia of *Aleochara***. For the section on this genus, I used Gack and Peschke (1994, 2005) and Putnam (1988). Claudia Gack read and approved the text. I recommend Godfray (1994) for general information on insect parasitoids.

125 The details of the act of **homosexual necrophilia** were taken from Moeliker (2001, 2009); Kees Moeliker also read and approved the text.

125–27 The papers by McCracken on **the Argentine lake duck** are McCracken (2000) and McCracken et al. (2001), whereas the quotes by Birkhead are from his Tinbergen Lecture in Leiden, May 11, 2012.

127 The information on **penis pecking** associated with duck mating on land is in Birkhead (2008: 313, 385).

127–28 The **work by Patricia Brennan** is in Brennan et al. (2007, 2009). The section in which the work by Birkhead and Brennan is mentioned was checked by Tim Birkhead. Parts of this paragraph are based on a Dutch-language article I wrote in *Bionieuws* (Schilthuizen, 2010).

128 Lange et al. (2013) is a general source of information on **traumatic insemination**.

128–30 The **copulation in *Harpactea sadistica*** is described in Řezáč (2009).

130–32 As a general source on **bedbug reproduction** I used Reinhardt and Siva-Jothy (2007). The quotes by Siva-Jothy are from his lecture at the Congress of the European Society for Evolutionary Biology, in Tübingen, Germany, August 20–24, 2011. Rivnay (1933) described how males detect moving females. The story that male bedbugs sometimes inseminate other males is in Lloyd (1979), but is considered not plausible by Judson (2002). Mellanby (1939) describes how females die after being housed with twelve males at the same time. The reduced life span caused by copulation is in Stutt and Siva-Jothy (2001). The experiments with replica bedbug penises are, for example, in Morrow and Arnqvist (2003).

133 The **sperm found in the blood of mites** was mentioned to me by Gerd Alberti and Antonella di Palma during an interview in Greifswald, Germany, on December 18, 2012; it is also in Alberti (2002).

133 The information on **insemination of birds and mammals by injection of sperm** is in Rowlands (1957).

133–34 The study of **pregnancy in women with noncommunicating uterine horns** is Nahum et al. (2004).

Chapter 7: Future Suitors

135–52 My **visit to Greifswald's Zoological Institute** and the interviews with Gerd Alberti, Antonella di Palma, Gabriele Uhl, Martin Haase, Peter Michalik, and Theo Schmitt took place December 17–19, 2012.

135–36 Basic **details on the reproduction of spiders and mites** are in Alberti (2002), Alberti and Coons (1999), and Alberti and Michalik (2004). The piece of text on my interview with Alberti and Di Palma was read and approved by the former.

136 An overview of **mating plugs in spiders** is in Uhl et al. (2010).

136–40 The **whole text on mating plugs in spiders** was read and approved by Gabriele Uhl.

137–39 The spread of the **wasp spider** is recorded in Kumschick et al. (2011), whereas the mating plug work in this species is in Nessler et al. (2007) and Uhl et al. (2007).

139–40 The ***Tidarren* work** is in Knoflach and Van Harten (2001), Knoflach and Benjamin (2003), and Michalik et al. (2010).

140–41 The **first observation of a mating plug** in a guinea pig is in Leuckart (1847; *fide* Dean, 2013), and in a chimpanzee in Tinklepaugh (1930).

140–45 I obtained information on **mating plugs in mammals** primarily from Kingan et al. (2003), Carnahan and Jensen-Seaman (2008), Tauber et al. (1975), Hernández-Lopez et al. (2008), Dean (2013), Murer et al. (2001), Koprowski (1992), and Dixson and Anderson (2002). I wrote about Kingan's work earlier (Schilthuizen, 2003).

145 Information on **women's allergic reaction to semen** is to be found at seminalplasmaallergy.org.

145–46 I got the information on **the numbers of proteins** in *Drosophila* ejaculate from Findlay et al. (2008) and in human ejaculate from Pilch and Mann (2006).

146 The **quote by Rama Singh** was taken from a lecture he gave at the ICSEB conference in Budapest, 1996; a recent paper by his group on this subject is Haerty et al. (2007).

146–47 The **way the *Drosophila* sex peptide works** is summarized in, for example, Eberhard (1996), Gillott (2003), and Liu and Kubli (2003).

147 The **mode of action of semen proteins in other insects** was taken from Gillott (2003) and Eberhard (1996). Specifically, the studies on fire beetles, corn earworm moths, and ticks are in Eisner et al. (1996), Kingan et al. (1995), and Leahy and Galun (1972), respectively.

147–48 Gallup et al. (2002) studied the **possible psychological effects of semen** in female college students.

148 An overview of the **relationship between semen and preeclampsia** is in Robertson et al. (2003).

148 The **grasshopper experiments** are from Hartmann and Loher (1999).

148–49 The **effects of seminal proteins on the female housefly** are in Leopold et al. (1971) and Riemann and Thorson (1969).

149 The two studies on **banana fly semen** that I mention are Chapman et al. (1995) and Civetta and Clark (2000).

149 The **seed beetle semen** is discussed in Eady et al. (2007) and Yamane and Miyatake (2012).

150 The paper on **spiny longhorn beetle genitalia** is Hubweber and Schmitt (2010).

151–52 Overviews of **spiny penises across the animal kingdom** are in Eberhard (1985) and Cordero and Miller (2012); the latter paper is also the source for the caltrop cornuti in moths. The penile spines of hoary bats can be admired in Cryan et al. (2012).

152 **I talked with Göran Arnqvist** on January 18, 2013, at the Linnaeusborg, University of Groningen, for the occasion of Bram Kuijper's Ph.D. defense.

152–53 **The article in *Nature*** is Crudgington and Siva-Jothy (2000).

153 **The postmating harm** inflicted by Arnqvist and colleagues is described in Morrow et al. (2003).

154 **The microlaser experiments** are in Hotzy et al. (2012).

154 **Other work on seed beetles by Arnqvist's team** is Rönn et al. (2007) and Hotzy and Arnqvist (2009).

154 **I spoke with Michal Polak** about his experimental setup (see Polak and Rashed, 2010) after his lecture at the Congress of the European Society for Evolutionary Biology in Tübingen, Germany, August 20–24, 2011.

154–55 General information on **primate penis spines** is from Zarrow and Clark (1968), Prasad (1970), and Stockley (2002), and for galagos in particular from Anderson (2000), Perkin (2007), and Veerman (2010). I also used the Web site of the Nocturnal Primate Research Group at Oxford Brookes University, UK.

155–56 The **vagina damage in rodents** with spiny penises is mentioned in Van der Schoot et al. (1992).

156 The experiments on **spine removal in marmosets** are in Dixson (1991).

156 The paper on the **genetics of human penis spinelessness** is McLean et al. (2011).

156 Kingsley's lab Web site on which he reports on the **papules vs. spines issue** is kingsley.stanford.edu/SpinesVsPapules.html.

Chapter 8: Sexual Ambivalence

158 The **lecture by Joris Koene** that I describe took place in Leiden on November 23, 2010.

158 The **mating behavior of *Deroceras praecox*** is described in Reise (2007) and Hutchinson and Reise (2009).

158 **The "condom" that some snails use** to prevent self-fertilization is described in Bojat et al. (2001).

159 The **taxonomic paper on *Deroceras praecox*** is Wiktor (1966).

160 The deposition of **secretions by the penial gland** is described (for a different *Deroceras* species) in Benke et al. (2010).

161 The **interview with Joris Koene** is constructed from two separate visits to his lab: once in 2008, and the second time on April 11, 2013.

161 The **role of Bateman's principle in hermaphrodites** is actually a more complex matter; see, for example, Leonard (2005).

162–63 The theory on **why hermaphrodites would be better at sexual manipulation** than organisms with separate sexes is in Koene (2005) and Michiels and Koene (2006).

163 **Penis fencing in flatworms** is described in Michiels and Newman (1998).

163 **General information on love darts** in snails is, for example, in Koene and Schulenburg (2005), Davison et al. (2005), Schilthuizen (2005), and Koene et al. (2013), in earthworms in Koene et al. (2002, 2005), and in sea slugs in Lange et al. (2012).

163 The **quotes** are from Jones (1841: 399).

164–67 The papers on the **true function of the love dart in *Cornu aspersum*** are Koene and Chase (1998a, b) and Chase and Blanchard (2006).

166–68 The **sexual arms race between dart evolution and countermeasures in the female genitalia** is in Koene and Schulenburg (2005).

168 **This and the previous section** were read and approved by Joris Koene.

168 The **quote on slugs and rats** is by Anthony Cook and was uttered during his lecture at the World Congress of Malacology in Siena, Italy, in 1992.

169–70 The **interviews with Martin Haase** were on December 17–19, 2012.

169–70 The **work on *Aeolidiella*** is described in Haase and Karlsson (2000, 2004) and Karlsson and Haase (2002). I wrote about this work earlier (Schilthuizen, 2001).

170–71 The **sea slug with disposable penis** is in Sekizawa et al. (2013) and Milius (2013).

171 The ***Deroceras* that amputates its own penis** is mentioned in Leonard (2006).

171 A summary of **penis-biting behavior in slugs** is in Reise and Hutchinson (2002). Penis chewing in banana slugs is in Leonard et al. (2002).

171–75 **The piece on *Limax*** is based on Lister (1678: 129–30), Redi (1684), Gerhardt (1933), and Glaubrecht and Zorn (2012), and material on the Web sites www.naturamediterraneo.com and www.wirbellose.at.

177 Some **general articles on coiling direction in snails** are Gittenberger (1988) and Schilthuizen and Davison (2005).

177–78 The **experiment by Meisenheimer** is in Meisenheimer (1912).

178–82 **Our work on *Amphidromus*** of the island of Kapas is described in, for example, Schilthuizen et al. (2005, 2007, 2009, 2012). It involved many collaborators, and I mention only one of them (Paul Craze) in the text, but in addition Lilian Wan, Sylvia Looijestijn, Sigrid Hendrikse, Kees Koops, Bronwen Scott, Annadel Cabanban, Martin Haase, Rachel Esner, and Angela Schmitz-Ornés have all helped, most of them in the field.

Afterplay

183 **Waage's memory of the press coverage** of his *Science* paper is in an e-mail correspondence I had with him on March 14, 2013.

183 The **original CNS story** leading to "Duckpenisgate" is at http://cnsnews.com/news/article/384949-federal-study-looks-plasticity-duck-penis-length. Carl Zimmer wrote about it on his blog *The Loom* at http://phenomena.nationalgeographic.com/2013/03/25/ducks-meet-the-culture-wars/. Patricia Brennan responded in an April 2, 2013, post in *Slate,* www.slate.com/articles/health_and_science/science/2013/04/duck_penis_controversy_nsf_is_right_to_fund_basic_research_that_conservatives.html.

184 **Eberhard's paper on artificial insemination** is Eberhard (1991).

184 A thoughtful essay on the **evolution of human sexual behavior** is Hrdy (1997).

185 The **castration of an alpha male** in Burgers' Zoo in Arnhem, the Netherlands, is described in De Waal (1986).

185 **Facebook's removal** of *Scientific American*'s post is mentioned on April 10, 2013, at slantist.com/facebook-censors-scientific-american/.

185–86 **The controversy over CFC versus SAC** is played out in, for example, Cordero and Eberhard (2003), Chapman et al. (2003), and Eberhard (2004a, b; 2010b). The Eberhard quotations are from an e-mail correspondence I had with him on April 25, 2013.

186 The **quote is from Gowaty** (1997: 353).

188 The work on **speciation in seed beetles** is in Brown and Eady (2001), Rönn et al. (2007), and Hotzy and Arnqvist (2009); more about the role of male-female coevolution in the splitting of species is in my book *Frogs, Flies, and Dandelions* (Schilthuizen, 2000).

188 **The role of homosexual behavior** in genital evolution in chimpanzees and bonobos is in Hrdy (1997). See also Bagemihl (2000) and Scharf & Martin (2013).

Bibliography

Alberti, G. 2002. "Ultrastructural Investigations of Sperm and Genital Systems in Gamasida (Acari: Anactinotrichida): Current State and Perspectives for Future Research." *Acarologia* 42:107–26.

Alberti, G., and L. B. Coons. 1999. "Acari—Mites." In *Microscopic Anatomy of Invertebrates,* vol. 8c, edited by F. W. Harrison and Rainer F. Foelix, 515–1265. New York: Wiley-Liss.

Alberti, G., and P. Michalik. 2004. "Feinstrukturelle Aspekte der Fortpflanzungssysteme von Spinnentieren (Arachnida)." *Denisia* 12:1–62.

Anderson, M. J. 2000. "Penile Morphology and Classification of Bush Babies (Subfamily Galagoninae)." *International Journal of Primatology* 21:815–36.

Araújo-Siqueira, M., and L. M. de Almeida. 2006. "Estudo das espécies brasileiras de *Cycloneda* Crotch (Coleoptera, Coccinellidae)." *Revista Brasileira de Zoologia* 23:550–68.

Arnqvist, G. 1997. "The Evolution of Animal Genitalia: Distinguishing Between Hypotheses by Single Species Studies." *Biological Journal of the Linnean Society* 60:365–79.

Bagemihl, B. 2000. *Biological Exuberance: Animal Homosexuality and Natural Diversity.* New York: Stonewall Inn.

Baker, R. R., and M. A. Bellis. 1993a. "Human Sperm Competition: Ejaculate Adjustment by Males and the Function of Masturbation." *Animal Behaviour* 46:861–85.

Baker, R. R., and M. A. Bellis. 1993b. "Human Sperm Competition: Ejaculate Manipulation by Females and a Function for the Female Orgasm." *Animal Behaviour* 46:887–909.

Baker, R. R., and M. A. Bellis. 1995. *Human Sperm Competition: Copulation, Masturbation and Infidelity.* London: Chapman and Hall.

Barker, D. M. 1994. "Copulatory Plugs and Paternity Assurance in the Nematode *Caenorhabditis elegans.*" *Animal Behaviour* 48:147–56.

Barlow, N. 1958. *The Autobiography of Charles Darwin, 1809–1882. With the Original Omissions Restored. Edited and with Appendix and Notes by his Grand-Daughter Nora Barlow.* London: Collins.

Barr, M. M., and L. R. Garcia. 2006. "Male Mating Behavior." In *WormBook,* edited by the *C. elegans* Research Community, doi:10.1895/wormbook.1.78.1, www.wormbook.org.

Barron, A. B., and M. J. F. Brown. 2012. "Science Journalism: Let's Talk About Sex." *Nature* 488:151–52.

Bassett, E. G. 1961. "Observations on the Retractor Clitoridis and Retractor Penis Muscles of Mammals, with Special Reference to the Ewe." *Journal of Anatomy* 95:61–77, pl. 1–3.

Bateman, A. J. 1948. "Intra-Sexual Selection in *Drosophila*." *Heredity* 2:349–68.

Baur, B. 2007. "Reproductive Biology and Mating Conflict in the Simultaneously Hermaphroditic Land Snail *Arianta arbustorum*." *American Malacological Bulletin* 23:157–72.

Baur, B. 2010. "Stylommatophoran Gastropods." In *The Evolution of Primary Sexual Characters in Animals,* edited by J. L. Leonard and A. Córdoba-Aguilar, 197–217. Oxford: Oxford University Press.

Benke, M., H. Reise, K. Montagne-Wajer, and J. M. Koene. 2010. "Cutaneous Application of an Accessory-Gland Secretion After Sperm Exchange in a Terrestrial Slug (Mollusca: Pulmonata)." *Zoology* 113:118–24.

Birkhead, T. R. 1995. "Human Sperm Competition: Copulation, Masturbation, and Infidelity" (book review). *Animal Behaviour* 50:1141–42.

Birkhead, T. R. 2007. "Promiscuity." *Daedalus* 136(2):13–22.

Birkhead, T. R. 2008. *The Wisdom of Birds: An Illustrated History of Ornithology.* London: Bloomsbury.

Birkhead, T. R. 2009. "Sex and Sensibility. *Times Higher Education Supplement,* February 5, 2009.

Birkhead, T. R., and A. P. Møller. 1993. "Sexual Selection and the Temporal Separation of Reproductive Events." *Biological Journal of the Linnean Society* 50:295–311.

Birkhead, T. R., H. D. M. Moore, and J. M. Bedford. 1997. "Sex, Science, and Sensationalism." *Trends in Ecology and Evolution* 12:121–22.

Blaicher, W., D. Gruber, C. Bieglmayer, A. M. Blaicher, W. Knogler, and J. C. Huber. 1999. "The Role of Oxytocin in Relation to Female Sexual Arousal." *Gynecologic and Obstetric Investigation* 47:125–26.

Bohme, W. 1983. "The Tucano Indians of Colombia and the Iguanid Lizard *Plica plica:* Ethnological, Herpetological and Ethological Implications." *Biotropica* 15:148–50.

Bojat, N. C., U. Sauder, and M. Haase. 2001. "The Spermathecal Epithelium, Sperm and Their Interactions in the Hermaphroditic Land Snail *Arianta arbustorum* (Pulmonata, Stylommatophora)." *Zoomorphology* 120:149–57.

Boomsma, J. J., B. Baer, and J. Heinze. 2005. "The Evolution of Male Traits in Social Insects." *Annual Review of Entomology* 50:395–420.

Boulangé, H. 1924. "Recherches sur l'appareil copulateur des Hymenoptères et spécialement des Chalastrogastres." *Mémoires et Travaux de la Faculté Catholique de l'Université de Lille* 28:1–444.

Brennan, P. L. R., C. J. Clark, and P. O. Prum. 2010. "Explosive Eversion and Functional Morphology of the Duck Penis Supports Sexual Conflict in Waterfowl Genitalia." *Proceedings of the Royal Society B* 277:1309–14.

Brennan, P. L. R., R. O. Prum, K. G. McCracken, M. D. Sorenson, R. E. Wilson, and T. R. Birkhead. 2007. "Coevolution of Male and Female Genital Morphology in Waterfowl." *PLOS ONE* 2:e418.

Brown, D. V., and P. E. Eady. 2001. "Functional Incompatibility Between the Fertilization Systems of Two Allopatric Populations of *Callosobruchus maculatus* (Coleoptera: Bruchidae)." *Evolution* 55:2257–62.

Bruce, H. M. 1959. "An Exteroceptive Block to Pregnancy in the Mouse." *Nature* 184:105.

Bruce, H. M. 1960. "A Block to Pregnancy in the Mouse Caused by Proximity of Strange Males." *Journal of Reproduction and Fertility* 1:96–103.

Bullini, L., and G. Nascetti. 1990. "Speciation by Hybridization in Phasmids and Other Insects." *Canadian Journal of Zoology* 68:1747–60.

Burger, M. 2007. "Sperm Dumping in a Haplogyne Spider." *Journal of Zoology* 273: 74–81.

Burger, M., W. Graber, P. Michalik, and C. Kropf. 2006. "*Silhouettella loricatula* (Arachnida, Araneae, Oonopidae): A Haplogyne Spider with Complex Female Genitalia." *Journal of Morphology* 267:663–77.

Burley, N. 1988. "The Differential-Allocation Hypothesis: An Experimental Test." *American Naturalist* 132:611–28.

Burley, N., G. Krantzberg, and P. Radman. 1982. "Influence of Color-Banding on the Conspecific Preferences of Zebra Finches." *Animal Behaviour* 30:444–55.

Burley, N. T., and R. Symanski. 1998. "'A Taste for the Beautiful': Latent Aesthetic Mate Preferences for White Crests in Two Species of Australian Grassfinches." *American Naturalist* 152:792–802.

Burns, E. 1953. *The Sex Life of Wild Animals: A North American Study.* New York: Rinehart.

Burton, F. D. 1971. "Sexual Climax in Female *Macaca mulatta*." In *Proceedings of the 3rd International Congress of Primatology, Zurich, 1970,* vol. 3:180–91.

Bussière, L. F., M. Demont, A. J. Pemberton, M. D. Hall, and P. I. Ward. 2010. "The Assessment of Insemination Success in Yellow Dung Flies Using Competitive PCR." *Molecular Ecology Resources* 10:292–303.

Butler, C. M., G. Shaw, and M. B. Renfree. 1999. "Development of the Penis and Clitoris in the Tammar Wallaby, *Macropus eugenii*." *Anatomy and Embryology* 199:451–57.

Calzada, J.-P. V., C. F. Crane, and D. M. Stelly. 1996. "Apomixis: The Asexual Revolution." *Science* 274:1322–23.

Carnahan, S. J., and M. I. Jensen-Seaman. 2008. "Hominoid Seminal Protein Evolution and Ancestral Mating Behavior." *American Journal of Primatology* 70:939–48.

Chapman, T., G. Arnqvist, J. Bangham, and L. Rowe. 2003. "Response to Eberhard and Cordero, and Córdoba-Aguilar and Contreras-Garduno: Sexual Conflict and Female Choice." *Trends in Ecology and Evolution* 18:440–41.

Chapman, T., L. F. Liddle, J. M. Kalb, M. F. Wolfner, and L. Partridge. 1995. "Cost of Mating in *Drosophila melanogaster* Females Is Mediated by Male Accessory Gland Products." *Nature* 373:241–44.

Chase, R., and K. C. Blanchard. 2006. "The Snail's Love-Dart Delivers Mucus to Increase Paternity." *Proceedings of the Royal Society B* 273:1471–75.

Civetta, A., and A. G. Clark. 2000. "Correlated Effects of Sperm Competition and Postmating Female Mortality." *Proceedings of the National Academy of Sciences* 97:13162–65.

Clark, W. C. 1981. "Sperm Transfer Mechanisms: Some Correlates and Consequences." *New Zealand Journal of Zoology* 8:49–65.

Cold, C. J., and K. A. McGrath. 1999. "Anatomy and Histology of the Penile and Clitoral Prepuce in Primates: An Evolutionary Perspective of the Specialised Sensory Tissue of the External Genitalia." In *Male and Female Circumcision,* edited by G. C. Denniston, F. M. Hodges, and M. F. Milos, 19–25. New York: Kluwer Academic/Plenum.

Conner, W. E. 1999. "'Un Chant d'Appel Amoureux': Acoustic Communication in Moths." *Journal of Experimental Biology* 202:1711–23.

Coope, G. R. 1979. "Late Cenozoic Fossil Coleoptera: Evolution, Biogeography, and Ecology." *Annual Review of Ecology and Systematics* 10:247–67.

Coope, G. R. 2004. "Several Million Years of Stability Among Insect Species Because of, or in Spite of, Ice Age Climatic Instability?" *Philosophical Transactions of the Royal Society B* 359:209–14.

Coope, G. R., and R. B. Angus. 1975. "An Ecological Study of a Temperate Interlude in the Middle of the Last Glaciation, Based on Fossil Coleoptera from Isleworth, Middlesex." *Journal of Animal Ecology* 44:365–91.

Cordero, C., and W. G. Eberhard. 2003. "Female Choice of Sexually Antagonistic Male Adaptations: A Critical Review of Some Current Research." *Journal of Evolutionary Biology* 16:1–6.

Cordero, C., and J. S. Miller. 2012. "On the Evolution and Function of Caltrop Cornuti in Lepidoptera—Potentially Damaging Male Genital Structures Transferred to Females During Copulation." *Journal of Natural History* 46:701–15.

Cordero-Rivera, A., and A. Córdoba-Aguilar. 2010. "Selective Forces Propelling Genitalic Evolution in Odonata." In *The Evolution of Primary Sexual Characters in Animals,* edited by J. L. Leonard and A. Córdoba-Aguilar, 332–52. Oxford: Oxford University Press.

Córdoba-Aguilar, A. 1999. "Male Copulatory Sensory Stimulation Induces Female Ejection of Rival Sperm in a Damselfly." *Proceedings of the Royal Society B* 266: 779–84.

Córdoba-Aguilar, A. 2002. "Sensory Trap as the Mechanism of Sexual Selection in a Damselfly Genitalic Trait (Insecta: Calopterygidae)." *American Naturalist* 160: 594–601.

Córdoba-Aguilar, A. 2005. "Possible Coevolution of Male and Female Genital Form and Function in a Calopterygid Damselfly." *Journal of Evolutionary Biology* 18: 132–37.

Córdoba-Aguilar, A., E. Uhia, and A. Cordero-Rivera. 2003. "Sperm Competition in Odonata (Insecta): The Evolution of Female Sperm Storage and Rivals' Sperm Displacement." *Journal of Zoology* 261:381–98.

Costa, R. M., G. F. Miller, and S. Brody. 2012. "Women Who Prefer Longer Penises Are More Likely to Have Vaginal Orgasms (But Not Clitoral Orgasms): Implications for an Evolutionary Theory of Vaginal Orgasm." *Journal of Sexual Medicine* 9:3079–88.

Cotton, S., J. Small, R. Hashim, and A. Pomiankowski. 2010. "Eyespan Reflects Reproductive Quality in Wild Stalk-Eyed Flies." *Evolutionary Ecology* 24:83–95.

Cryan, P. M., J. W. Jameson, E. F. Baerwald, C. K. R. Willis, R. M. R. Barclay, E. A. Snider, and E. G. Crichton. 2012. "Evidence of Late-Summer Mating Readiness and Early Sexual Maturation in Migratory Tree-Roosting Bats Found Dead at Wind Turbines." *PLOS ONE* 7:e47586.

Crudgington, H. S., and M. T. Siva-Jothy. 2000. "Genital Damage, Kicking, and Early Death." *Nature* 407:855–56.

Cuvier, G. 1829. *Iconographie du Règne Animal; ou, Représentation d'après Nature de l'une des espèces les plus et souvent non encore figurées de chaque genre d'animaux.* Paris: Baillière.

D'Aguilar, K. 2007. "René Jeannel, l'homme des cavernicoles." *Insectes* 146:31–32.

Dahl, J. F. 1985. "The External Genitalia of Female Pygmy Chimpanzees." *Anatomical Record* 211:24–28.

Darwin, C. R. 1851. *A Monograph of the Sub-Class Cirripedia, with Figures of all the Species. The Lepadidae; or, the Pedunculated Cirripedes.* London: Ray Society.

Darwin, C. R. 1854. *A Monograph of the Sub-Class Cirripedia, with Figures of all the Species. The Balanidae (or Sessile Cirripedes); the Verrucidae, etc.* London: Ray Society.

Darwin, C. R. 1859. *On the Origin of Species by Means of Natural Selection, or The Preservation of Favoured Races in the Struggle for Life.* London: John Murray.

Darwin, C. R. 1871. *The Descent of Man, and Selection in Relation to Sex.* London: John Murray.

Davison, A., C. M. Wade, P. B. Mordan, and S. Chiba. 2005. "Sex and Darts in Slugs and Snails (Mollusca: Gastropoda: Stylommatophora)." *Journal of Zoology* 267:329–38.

Dawkins, R. 2004. *The Ancestor's Tale: A Pilgrimage to the Dawn of Life.* London: Weidenfeld and Nicolson.

Dawkins, R., and J. R. Krebs. 1979. "Arms Races Between and Within Species." *Proceedings of the Royal Society B* 205:489–511.

Dean, M. D. 2013. "Genetic Disruption of the Copulatory Plug in Mice Leads to Severely Reduced Fertility." *PLOS Genetics* 9:e1003185.

Dean, R., S. Nakagawa, and T. Pizzari. 2011. "The Risk and Intensity of Sperm Ejection in Female Birds." *American Naturalist* 178:343–54.

De Wilde, J. 1964. "Reproduction-Endocrine Control." In *The Physiology of Insecta*, vol. 1, edited by M. Rockstein, 59–90. New York: Academic Press.

Dewsbury, D. A. 1971. "Copulatory Behaviour of Old-Field Mice (*Peromyscus polionotus subgriseus*)." *Animal Behaviour* 19:192–204.

Dewsbury, D. A. 1974. "Copulatory Behavior of California Mice (*Peromyscus californicus*)." *Brain, Behavior, and Evolution* 9:95–106.

Dixson, A. F. 1991. "Penile Spines Affect Copulatory Behaviour in a Primate (*Callithrix jacchus*)." *Physiology and Behavior* 49:557–62.

Dixson, A. F., and M. J. Anderson. 2002. "Sexual Selection, Seminal Coagulation and Copulatory Plug Formation in Primates." *Folia Primatologica* 73:63–69.

Düngelhoef, S., and M. Schmitt. 2010. "Genital Feelers: The Putative Role of Parameres and Aedeagal Sensilla in Coleoptera Phytophaga (Insecta)." *Genetica* 138: 45–57.

Dybas, H. S. 1978. "The Systematics, Geographical and Ecological Distribution of *Ptiliopycna*, a Nearctic Genus of Parthenogenetic Featherwing Beetles (Coleoptera: Ptiliidae)." *American Midland Naturalist* 99:83–100.

Eady, P. E., I. Hamilton, and R. E. Lyons. 2007. "Copulation, Genital Damage and Early Death in *Callosobruchus maculatus*." *Proceedings of the Royal Society B* 272:247–52.

Eberhard, W. G. 1985. *Sexual Selection and Animal Genitalia*. Cambridge, MA: Harvard University Press.

Eberhard, W. G. 1991. "Artificial Insemination: Can Appropriate Stimulation Improve Success Rates?" *Medical Hypotheses* 36:152–54.

Eberhard, W. G. 1996. *Female Control: Sexual Selection by Cryptic Female Choice*. Princeton, NJ: Princeton University Press.

Eberhard, W. G. 2004a. "Rapid Divergent Evolution of Sexual Morphology: Comparative Tests of Antagonistic Coevolution and Traditional Female Choice." *Evolution* 58:1947–70.

Eberhard, W. G. 2004b. "Male-Female Conflict and Genitalia: Failure to Confirm Predictions in Insects and Spiders." *Biological Reviews* 79:121–86.

Eberhard, W. G. 2008. "Static Allometry and Animal Genitalia." *Evolution* 63:48–66.

Eberhard, W. G. 2010a. "Rapid Divergent Evolution of Genitalia: Theory and Data Updated." In *The Evolution of Primary Sexual Characters in Animals*, edited by J. L. Leonard and A. Córdoba-Aguilar, 40–78. Oxford: Oxford University Press.

Eberhard, W. G. 2010b. "Evolution of Genitalia: Theory, Evidence, and New Directions." *Genetica* 138:5–18.

Eberhard, W. G., and J. K. Gelhaus. 2009. "Genitalic Stridulation During Copulation in a Species of Crane Fly, *Tipula* (*Bellardina*) sp. (Diptera: Tipulidae)." *International Journal of Tropical Biology* 57 (Suppl. 1):251–56.

Eberhard, W. G., and B. A. Huber. 2010. "Spider Genitalia: Precise Maneuvers with a Numb Structure in a Complex Lock." In *The Evolution of Primary Sexual Characters in Animals*, J. L. Leonard and A. Córdoba-Aguilar, 249–84. Oxford: Oxford University Press.

Eberhard, W. G., B. A. Huber, R. L. Rodriguez S., R. D. Briceno, I. Salas, and V. Rodriguez. 1998. "One Size Fits All? Relationships Between the Size and Degree of Variation in Genitalia and Other Body Parts in Twenty Species of Insects and Spiders." *Evolution* 52:415–31.

Eilperin, J. 2012. *Demon Fish: Travels Through the Hidden World of Sharks*. New York: Anchor.

Eisner, T., S. R. Smedley, D. K. Young, M. Eisner, B. Roach, and J. Meinwald. 1996. "Chemical Basis of Courtship in a Beetle (*Neopyrochroa flabellata*): Cantharidin as 'Nuptial Gift.' *Proceedings of the National Academy of Sciences* 93:6499–503.

Eldredge, N., and S. J. Gould. 1972. "Punctuated Equilibria: An Alternative to Phyletic Gradualism." In *Models in Paleobiology*, edited by T. J. M. Schopf, 82–115. San Francisco: Freeman, Cooper.

Endler, J. A., and A. L. Basolo. 1998. "Sensory Ecology, Receiver Biases and Sexual Selection." *Trends in Ecology and Evolution* 13:415–20.

Evanno, G., and L. Madec. 2007. "Variation morphologique de la spermathèque chez l'escargot terrestre *Cantareus aspersus.*" *Comptes Rendus Biologies* 330:722–27.

Evans, L. E., and D. S. McKenna. 1986. "Artificial Insemination of Swine." In *Current Therapy in Theriogenology,* edited by D. Morrow, 946–49. Philadelphia: W. B. Saunders.

Findlay, G. D., X. Yi, M. J. MacCoss, and W. J. Swanson. 2008. "Proteomics Reveals Novel *Drosophila* Seminal Fluid Proteins Transferred at Mating." *PLOS Biology* 6:e178.

Fitzpatrick, J. L., R. M. Kempster, T. S. Daly-Engel, S. P. Collin, and J. P. Evans. 2012. "Assessing the Potential for Post-Copulatory Sexual Selection in Elasmobranchs." *Journal of Fish Biology* 80:1141–58.

Foldes, P., and O. Buisson. 2009. "The Clitoral Complex: A Dynamic Sonographic Study." *Journal of Sexual Medicine* 6:1223–31.

Fontaneto, D., E. A. Herniou, C. Boscheti, M. Caprioli, G. Melone, C. Ricci, and T. G. Barraclough. 2007. "Independently Evolving Species in Asexual Bdelloid Rotifers." *PLOS Biology* 5:e87.

Fooden, J. 1967. "Complementary Specialization of Male and Female Reproductive Structures in the Bear Macaque, *Macaca arctoides.*" *Nature* 214:939–41.

Forbes, L. S. 1997. "The Evolutionary Biology of Spontaneous Abortion in Humans." *Trends in Ecology and Evolution* 12:446–50.

Fox, C. A., H. S. Wolff, and J. A. Baker. 1970. "Measurement of Intra-Vaginal and Intra-Uterine Pressures During Human Coitus by Radio-Telemetry." *Journal of Reproduction and Fertility* 22:243–51.

Freud, S. 1905. *Drei Abhandlungen zur Sexualtheorie.* Vienna: Deuticke.

Freude, H., K. Harde, and G. Lohse. 1971. *Die Käfer Mitteleuropas, Band 3: Adephaga II (Hygrobiidae—Rhysodidae), Palpicornia (Hydraenidae—Hydrophilidae), Histeroidea, Staphylinoidea (exkl. Staphylinidae).* Krefeld, Germany: Goecke and Evers.

Gack, C., and K. Peschke. 1994. "Spermathecal Morphology, Sperm Transfer and a Novel Mechanism of Sperm Displacement in the Rove Beetle, *Aleochara curtula* (Coleoptera, Staphylinidae)." *Zoomorphology* 114:227–37.

Gack, C., and K. Peschke. 2005. "'Shouldering' Exaggerated Genitalia: A Unique Behavioural Adaptation for the Retraction of the Elongate Intromittant Organ by the Male Rove Beetle (*Aleochara tristis* Gravenhorst)." *Biological Journal of the Linnean Society* 84:307–12.

Gallup, G. G., R. L. Burch, and T. J. B. Mitchell. 2006. "Semen Displacement as a Sperm Competition Strategy." *Human Nature* 17:253–64.

Gallup, G. G., R. L. Burch, and S. M. Platek. 2002. "Does Semen Have Antidepressant Properties?" *Archives of Sexual Behavior* 31:289–93.

Gallup, G. G., R. L. Burch, M. L. Zappieri, R. A. Parvez, M. L. Stockwell, and J. A. Davis. 2003. "The Human Penis as a Semen Displacement Device." *Evolution and Human Behavior* 24:277–89.

Georgiadis, J. R. 2011. "Exposing Orgasm in the Brain: A Critical Eye." *Sexual and Relationship Therapy* 26:342–55.

Gerhardt, U. 1933. "Zur Kopulation der Limaciden. I. Mitteilung." *Zeitschrift für Morphologie und Ökologie der Tiere* 27:401–50.

Gerlach, N. M., J. W. McGlothlin, P. G. Parker, and E. D. Ketterson. 2012. "Reinterpreting Bateman Gradients: Multiple Mating and Selection in Both Sexes of a Songbird Species." *Behavioral Ecology* 23:1078–88.

Ghiselin, M. T. 1984. "'Definition,' 'Character,' and Other Equivocal Terms." *Systematic Zoology* 33:104–10.

Ghiselin, M. T. 2010. "The Distinction Between Primary and Secondary Sexual Characters." In *The Evolution of Primary Sexual Characters in Animals*, edited by J. L. Leonard and A. Córdoba-Aguilar, 9–14. Oxford: Oxford University Press.

Gilbert, P. W., and G. W. Heath. 1972. "The Clasper-Siphon Sac Mechanism in *Squalus acanthias* and *Mustelus canis*." *Comparative Biochemistry and Physiology* 42A:97–119.

Gillott, C. 2003. "Male Accessory Gland Secretions: Modulators of Female Reproductive Physiology and Behavior." *Annual Review of Entomology* 48:163–84.

Ginsberg, J. R., and U. W. Huck. 1989. "Sperm Competition in Mammals." *Trends in Ecology and Evolution* 4:74–79.

Ginsberg, J. R., and D. I. Rubenstein. 1990. "Sperm Competition and Variation in Zebra Mating Behavior." *Behavioral Ecology and Sociobiology* 26:427–34.

Giraldi, A., L. Marson, R. Nappi, J. Pfaus, A. M. Traish, Y. Vardi, and I. Goldstein. 2004. "Physiology of Female Sexual Function: Animal Models." *Journal of Sexual Medicine* 1:237–52.

Gist, D. H., and J. M. Jones. 1989. "Sperm Storage Within the Oviduct of Turtles." *Journal of Morphology* 199:379–84.

Gittenberger, E. 1988. "Sympatric Speciation in Snails: A Largely Neglected Model." *Evolution* 42:826–28.

Glaubrecht, M., and C. Zorn. 2012. "More Slug(gish) Science: Another Annotated Catalogue on Types of Tropical Pulmonate Slugs (Mollusca, Gastropoda) in the Collection of the Natural History Museum Berlin." *Zoosystematics and Evolution* 88:33–51.

Godfray, H. C. J. 1994. *Parasitoids: Behavioral and Evolutionary Ecology.* Princeton, NJ: Princeton University Press.

Gosse, P. H. 1883. "On the Clasping-Organs Ancillary to Generation in Certain Groups of the Lepidoptera." *Transactions of the Linnean Society of London (Zoology)* 2:265–345.

Gould, S. J. 1991. *Bully for Brontosaurus: Reflections in Natural History.* New York: Norton.

Gowaty, P. A. 1997. *Feminism and Evolutionary Biology: Boundaries, Intersections and Frontiers.* Heidelberg, Germany: Springer.

Gowaty, P. A., Y.-K. Kim, and W. W. Anderson. 2012. "No Evidence of Sexual Selection in a Repetition of Bateman's Classic Study of *Drosophila melanogaster.*" *Proceedings of the National Academy of Sciences* 109:11740–45.

Gruenwald, I., L. Lowenstein, I. Gartman, and Y. Vardi. 2007. "Physiological Changes in Female Genital Sensation During Sexual Stimulation." *Journal of Sexual Medicine* 4:390–94.

Haase, M., and A. Karlsson. 2000. "Mating and the Inferred Function of the Genital System of the Nudibranch, *Aeolidiella glauca* (Gastropoda: Opisthobranchia: Aeolidiodea)." *Invertebrate Biology* 119:287–98.

Haase, M., and A. Karlsson. 2004. "Mate Choice in a Hermaphrodite: You Won't Score with a Spermatophore." *Animal Behaviour* 67:287–91.

Haeberle, E. J. 1983. *The Sex Atlas: New Popular Reference Edition.* New York: Continuum.

Haerty, W., S. Jagadeeshan, R. J. Kulathinal, A. Wong, K. R. Ram, L. K. Sirot, L. Levesque, C. G. Artieri, M. F. Wolfner, A. Civetta, and R. S. Singh. 2007. "Evolution in the Fast Lane: Rapidly Evolving Sex-Related Genes in *Drosophila.*" *Genetics* 177:1321–35.

Hanby, J. P., and C. E. Brown. 1974. "The Development of Sociosexual Behaviors in Japanese Macaques *Macaca fuscata.*" *Behaviour* 49:152–95.

Harrison, P. L., R. C. Babcock, G. D. Bull, J. K. Oliver, C. C. Wallace, and B. L. Willis. 1984. "Mass Spawning in Tropical Coral Reefs." *Science* 223:1186–89.

Harrisson, T. 1959. *World Within: A Borneo Story.* London: Cresset Press.

Hartmann, R., and W. Loher. 1999. "Post-Mating Effects in the Grasshopper, *Gomphocerus rufus* L. Mediated by the Spermatheca." *Journal of Comparative Physiology A* 184:325–32.

Haubruge, E., L. Arnaud, J. Mignon, and M. J. G. Gage. 1999. "Fertilization by Proxy: Rival Sperm Removal and Translocation in a Beetle." *Proceedings of the Royal Society B* 266:1183–87.

Heller, K.-G., and R. Krahe. 1994. "Sound Production and Hearing in the Pyralid Moth *Symmoracma minoralis.*" *Journal of Experimental Biology* 187:101–11.

Hellriegel, B., and G. Bernasconi. 2000. "Female-Mediated Differential Sperm Storage in a Fly with Complex Spermathecae, *Scatophaga stercoraria.*" *Animal Behaviour* 59:311–17.

Helsdingen, P. J. van. 1965. "Sexual Behaviour of *Lepthyphantes leprosus* (Ohlert) (Araneida, Linyphiidae), with Notes on the Function of the Genital Organs." *Zoologische Mededelingen* (Leiden) 41:15–42.

Helversen, D. von, and O. von Helversen. 1991. "Pre-Mating Sperm Removal in the Bushcricket *Metaplastes ornatus* Ramme 1931 (Orthoptera, Tettigonoidea, Phaneropteridae)." *Behavioral Ecology and Sociobiology* 28:391–96.

Hernández-Lopez, L., A. L. Cerda-Molina, D. L. Páez-Ponce, and R. Mondragón-Ceballos. 2008. "The Seminal Coagulum Favours Passage of Fast-Moving Sperm into the Uterus in the Black-Handed Spider Monkey." *Reproduction* 136: 411–21.

Highnam, K. C. 1964. *Insect Reproduction.* London: Royal Entomological Society.

Hoch, J. M. 2008. "Variation in Penis Morphology and Mating Ability in the Acorn Barnacle, *Semibalanus balanoides.*" *Journal of Experimental Marine Biology and Ecology* 359:126–30.

Hoekstra, R. F. 1987. "The Evolution of Sexes." In *The Evolution of Sex and Its Consequences,* edited by S. C. Stearns, 59–91. Basel, Switzerland: Birkhäuser.

Holt, W. V., and R. E. Lloyd. 2010. "Sperm Storage in the Vertebrate Female Reproductive Tract: How Does It Work So Well?" *Theriogenology* 73:713–22.

Holwell, G. I., and M. E. Herberstein. 2010. "Chirally Dimorphic Male Genitalia in Praying Mantids (*Ciulfina:* Liturgusidae)." *Journal of Morphology* 271:1176–84.

Hormiga, G., and N. Scharff. 2005. "Monophyly and Phylogenetic Placement of the Spider Genus *Labulla* Simon, 1884 (Araneae, Linyphiidae) and Description of the New Genus *Pecado*." *Zoological Journal of the Linnean Society* 143: 359–404.

Hormiga, G., N. Scharff, and J. A. Coddington. 2000. "The Phylogenetic Basis of Sexual Size Dimorphism in Orb-Weaving Spiders (Araneae, Oribiculariae)." *Systematic Biology* 49:435–62.

Hosken, D. J. 2008. "Clitoral Variation Says Nothing About Female Orgasm." *Evolution and Development* 10:393–95.

Hotzy, C., and G. Arnqvist. 2009. "Sperm Competition Favors Harmful Males in Seed Beetles." *Current Biology* 19:404–7.

Hotzy, C., M. Polak, J. L. Rönn, and G. Arnqvist. 2012. "Phenotypic Engineering Unveils the Function of Genital Morphology." *Current Biology* 22:2258–61.

Houck, L. D., and P. A. Verrell. 2010. "Evolution of Primary Sexual Characters in Amphibians." In *The Evolution of Primary Sexual Characters in Animals*, edited by J. L. Leonard and A. Córdoba-Aguilar, 409–24. Oxford: Oxford University Press.

Hrdy, S. B. 1997. "Raising Darwin's Consciousness: Female Sexuality and the Prehominid Origins of Patriarchy." *Human Nature* 8:1–49.

Hubweber, L., and M. Schmitt. 2010. "Differences in Genitalia Structure and Function Between Subfamilies of Longhorn Beetles (Coleoptera: Cerambycidae)." *Genetica* 138:37–43.

Hughes, K. A., L. Du, H. Rodd, and D. N. Reznick. 1999. "Familiarity Leads to Female Mate Preference for Novel Males in the Guppy, *Poecilia reticulata*." *Animal Behaviour* 58:907–16.

Hunter, J. 1780. "Account of an Extraordinary Pheasant." *Philosophical Transactions of the Royal Society of London* 70:527–35.

Hutchinson, J. M. C., and H. Reise. 2009. "Mating Behaviour Clarifies the Taxonomy of Slug Species Defined by Genital Anatomy: The *Deroceras rodnae* Complex in the Sächsische Schweiz and Elsewhere." *Mollusca* 27:183–200.

Iwasa, Y., and A. Pomiankowski. 1995. "Continual Change in Mate Preferences." *Nature* 377:420–22.

Jann, P., W. U. Blanckenhorn, and P. I. Ward. 2000. "Temporal and Microspatial Variation in the Intensities of Natural and Sexual Selection in the Yellow Dung Fly *Scathophaga stercoraria*." *Journal of Evolutionary Biology* 13:927–38.

Jeannel, R. 1911. "Biospeologica 19: Révision des Bathysciinae (Coléoptères Silphides): morphologie, distribution géographique, systématique." *Archives de Zoologie Expérimentale et Générale 5, Série* 7:1–641, pl. i–xxiv.

Jeannel, R. 1936. "Monographie des Catopidae." *Mémoires du Muséum National d'Histoire Naturelle*, N. S. 1:1–435.

Jeannel, R. 1941. "L'isolement, facteur de l'évolution." *Revue Française d'Entomologie* 8:101–10.

Jeannel, R. 1955. *L'Édéage: initiation aux recherches sur la systématique des Coléoptères*. Paris: Publications du Muséum National d'Histoire Naturelle.

Jeyarajasingam, A., and A. Pearson. 1999. *A Field Guide to the Birds of West Malaysia and Singapore*. Oxford: Oxford University Press.

Jones, S. 2008. "A Wonderful Life by Leaps and Bounds." *Nature* 456:873–74.

Jones, T. R. 1841. *A General Outline of the Organisation of the Animal Kingdom, and Manual of Comparative Anatomy.* London: John van Voorst.

Joyce, W. G., N. Micklich, S. F. K. Schaal, and T. M. Scheyer. 2012. "Caught in the Act: The First Record of Copulating Fossil Vertebrates." *Biology Letters* 8:846–48.

Judson, O. P. 2002. *Dr. Tatiana's Sex Advice to All Creation.* London: Random House.

Judson, O. P. 2005. "Anticlimax." *Nature* 436:916–17.

Kahn, A. T., B. Mautz, and M. D. Jennions. 2010. "Females Prefer to Associate with Males with Longer Intromittent Organs in Mosquitofish." *Biology Letters* 6: 55–58.

Kamimura, Y., and H. Mitsumoto. "Lock-and-Key Structural Isolation Between Sibling *Drosophila* Species." *Entomological Science* 15:197–201.

Karlsson, A., and M. Haase. 2002. "The Enigmatic Mating Behaviour and Reproduction of a Simultaneous Hermaphrodite, the Nudibranch *Aeolidiella glauca* (Gastropoda, Opisthobranchia)." *Canadian Journal of Zoology* 80:260–70.

Karlsson Green, K., and J. A. Madjidian. 2011. "Active Males, Reactive Females: Stereotypic Sex Roles in Sexual Conflict Research?" *Animal Behaviour* 81:901–7.

Kelly, C. D., J.-G. J. Godin, and G. Abdallah. 2000. "Geographical Variation in the Male Intromittent Organ of the Trinidadian Guppy (*Poecilia reticulata*)." *Canadian Journal of Zoology* 78:1674–80.

Kingan, S. B., M. Tatar, and D. M. Rand. 2003. "Reduced Polymorphism in the Chimpanzee Semen Coagulating Protein, Semenogelin I." *Journal of Molecular Evolution* 57:159–69.

Kingan, T. G., W. M. Bodnar, A. K. Raina, J. Shabanowitz, and D. F. Hunt. 1995. "The Loss of Female Sex Pheromone After Mating in the Corn Earworm Moth *Helicoverpa zea:* Identification of a Male Pheromonostatic Peptide." *Proceedings of the National Academy of Sciences* 92:5082–86.

Kleukers, R. M. J. C., E. J. van Nieukerken, B. Odé, L. P. M. Willemse, and W. K. R. E. van Wingerden. 1997. *De Sprinkhanen en Krekels van Nederland (Orthoptera).* Leiden, the Netherlands: Nationaal Natuurhistorisch Museum, KNNV Uitgeverij and EIS-Nederland.

Klingenberg, C. P. 1996. "Multivariate Allometry." In *Advances in Morphometrics,* edited by L. F. Markus, M. Corti, A. Loy, G. J. P. Naylor, and D. E. Slice, 23–49. Heidelberg, Germany: Springer.

Knapp, L. A., J. C. Ha, and J. P. Sackett. 1996. "Parental MHC Antigen Sharing and Pregnancy Wastage in Captive Pigtailed Macaques." *Journal of Reproductive Immunology* 32:73–88.

Knoflach, B., and A. van Harten. 2000. "Palpal Loss, Single Palp Copulation and Obligatory Mate Consumption in *Tidarren cuneolatum* (Tullgren, 1910) (Araneae, Theridiidae)." *Journal of Natural History* 34:1639–59.

Knoflach, B., and S. P. Benjamin. 2003. "Mating Without Sexual Cannibalism in *Tidarren sisyphoides* (Araneae, Theridiidae)." *Journal of Arachnology* 31: 445–48.

Kobelt, G. L. 1844. *Die männlichen und weiblichen Wollust-Organe des Menschen und einiger Säugetiere.* Freiburg: n.p.

Kočárek, P. 2002. "Diel Activity Patterns of Carrion-Visiting Coleoptera Studied by Time-Sorting Pitfall Traps." *Biologia* (Bratislava) 57:199–211.

Koene, J. M. 2005. "Allohormones and Sensory Traps: A Fundamental Difference Between Hermaphrodites and Gonochorists?" *Invertebrate Reproduction and Development* 48:101–7.

Koene, J. M., and R. Chase. 1998a. "The Love Dart of *Helix aspersa* Müller Is Not a Gift of Calcium." *Journal of Molluscan Studies* 64:75–80.

Koene, J. M., and R. Chase. 1998b. "Changes in the Reproductive System of the Snail *Helix aspersa* Caused by Mucus from the Love Dart." *Journal of Experimental Biology* 201:2313–19.

Koene, J. M., T.-S. Liew, K. Montagne-Wajer, and M. Schilthuizen. 2013. "A Syringe-Like Love Dart Injects Male Accessory Gland Products in a Tropical Hermaphrodite." *PLOS ONE* 8:e69968.

Koene, J. M., T. Pförtner, and N. K. Michiels. 2005. "Piercing the Partner's Skin Influences Sperm Uptake in the Earthworm *Lumbricus terrestris*." *Behavioral Ecology and Sociobiology* 59:243–49.

Koene, J. M., and H. Schulenburg. 2005. "Shooting Darts: Co-Evolution and Counter-Adaptation in Hermaphroditic Snails." *BMC Evolutionary Biology* 5:e25.

Koene, J. M., G. Sundermann, and N. K. Michiels. 2002. "On the Function of Body Piercing During Copulation in Earthworms." *Invertebrate Reproduction and Development* 41:35–40.

Kokko, H., R. Brooks, M. D. Jennions, and J. Morley. 2003. "The Evolution of Mate Choice and Mating Biases." *Proceedings of the Royal Society B* 270:653–64.

Kokko, H., M. D. Jennions, and A. Houde. 2007. "Evolution of Frequency-Dependent Mate Choice: Keeping Up with Fashion Trends." *Proceedings of the Royal Society B* 274:1317–24.

Koprowski, J. L. 1992. "Removal of Copulatory Plugs by Female Tree Squirrels." *Journal of Mammalogy* 73:572–76.

Kuijper, B., I. Pen, and F. J. Weissing. 2012. "A Guide to Sexual Selection Theory." *Annual Review of Ecology, Evolution and Systematics* 43:287–311.

Kullmann, E. 1964. "Neue Ergebnisse über den Netzbau und das Sexualverhalten einiger Spinnenarten." *Zeitschrift für Zoologische Systematik und Evolutionsforschung* 2:41–122.

Kumschick, S., S. Fronzek, M. H. Entling, and W. Nentwig. 2011. "Rapid Spread of the Wasp Spider *Argiope bruennichi* Across Europe: A Consequence of Climate Change?" *Climate Change* 109:319–29.

Lange, R., T. Gerlach, J. Beninde, J. Werminghausen, V. Reichel, and N. Anthes. 2012. "Female Fitness Optimum at Intermediate Mating Rates Under Traumatic Mating." *PLOS ONE* 7:e43234.

Lange, R., K. Reinhardt, N. K. Michiels, and N. Anthes. 2013. "Functions, Diversity, and Evolution of Traumatic Mating. *Biological Reviews* 88:585–601.

Langerhans, R. B., C. A. Layman, and T. J. DeWitt. 2005. "Male Genital Size Reflects a Tradeoff Between Attracting Males and Avoiding Predators in Two Live-Bearing Fish Species." *Proceedings of the National Academy of Sciences* 102: 7618–23.

Langtimm, C. A., and D. A. Dewsbury. 1991. "Phylogeny and Evolution of Rodent Copulatory Behaviour." *Animal Behaviour* 41:217–24.

Laptikhovsky, V. V., and A. Salman. 2003. "On Reproductive Strategies of the Epipelagic Octopods of the Superfamily Argonautoidea (Cephalopoda: Octopoda)." *Marine Biology* 142:321–26.

Lawson, A. J. 2012. *Painted Caves: Palaeolithic Rock Art in Western Europe.* Oxford: Oxford University Press.

Leahy, M. G., and R. Galun. 1972. "Effect of Mating on Oogenesis and Oviposition in the Tick *Argas persicus* (Oken)." *Parasitology* 65:167–78.

Leeuwenhoek, A. van. 1678. "Observationes D. Anthonii Lewenhoeck de natis e semini genitali animalculis." *Philosophical Transactions of the Royal Society of London* 12:1040–46.

Leonard, J. L. 2005. "Bateman's Principle and Simultaneous Hermaphrodites: A Paradox." *Integrative and Comparative Biology* 45:856–73.

Leonard, J. L. 2006. "Sexual Selection: Lessons from Hermaphrodite Mating Systems." *Integrative and Comparative Biology* 46:349–67.

Leonard, J. L., J. S. Pearse, and A. B. Harper. 2002. "Comparative Reproductive Biology of *Ariolimax californicus* and *A. dolichophallus* (Gastropoda: Stylommatophora)." *Invertebrate Reproduction and Development* 41:83–93.

Leopold, R. A., A. C. Terranova, B. J. Thorson, and M. E. Degrugillier. 1971. "The Biosynthesis of the Male Housefly Accessory Secretion and Its Fate in the Mated Female." *Journal of Insect Physiology* 17:987–1003.

Leuckart, R. 1847. "Zur Morphologie und Anatomie der Geschlechtsorgane." Göttingen, Germany: Vandenhoeck und Ruprecht.

Lindroth, A. 1947. "Time of Activity of Freshwater Fish Spermatozoa in Relation to Temperature." *Zoologiska Bidrag fran Uppsala* 25:165–68.

Lister, M. 1678. *Historia Animalium Angliae tres tractatus.* London.

Litchfield, R. B. 1922. "Charles Darwin's Death-Bed: Story of Conversion Denied." *The Christian,* February 23, 1922, 12.

Liu, H., and E. Kubli. 2003. "Sex-Peptide Is the Molecular Basis of the Sperm Effect in *Drosophila melanogaster.*" *Proceedings of the National Academy of Sciences* 100:9929–33.

Lloyd, E. A. 2005. *The Case of the Female Orgasm: Bias in the Science of Evolution.* Cambridge, MA: Harvard University Press.

Lloyd, J. E. 1979. "Mating Behavior and Natural Selection." *Florida Entomologist* 62:17–34.

Long, J. A. 2012. *The Dawn of the Deed: The Prehistoric Origins of Sex.* Chicago: University of Chicago Press.

Long, J. A., K. Trinajstic, G. C. Young, and T. Senden. 2008. "Live Birth in the Devonian Period." *Nature* 453:650–52.

Long, J. A., K. Trinajstic, and Z. Johanson. 2009. "Devonian Arthrodire Embryos and the Origin of Internal Fertilization in Vertebrates." *Nature* 457:1124–27.

Lynch, V. J. 2008. "Clitoral and Penile Size Variability Are Not Significantly Different: Lack of Evidence for the Byproduct Theory of the Female Orgasm." *Evolution and Development* 10:396–97.

Maan, M. E., and M. E. Cummings. 2008. "Female Preferences for Aposematic Signal Components in a Polymorphic Poison Frog." *Evolution* 62:2334–45.

Maan, M. E., and M. E. Cummings. 2009. "Sexual Dimorphism and Directional Sexual Selection on Aposematic Signals in a Poison Frog." *Proceedings of the National Academy of Sciences* 106:19072–77.

Maan, M. E., and M. E. Cummings. 2012. "Poison Frog Colors Are Honest Signals of Toxicity, Particularly for Bird Predators." *American Naturalist* 179:e1–e14.

Marian, J. E. A. R., Y. Shiraki, K. Kawai, S. Kojima, Y. Suzuki, and K. Ono. 2012. "Revisiting a Medical Case of 'Stinging' in the Human Oral Cavity Caused by Ingestion of Raw Squid (Cephalopoda: Teuthida): New Data on the Functioning of Squid's Spermatophores." *Zoomorphology* 131:293–301.

Marson, L., R. Cai, and N. Makhanova. 2003. "Identification of Spinal Neurons Involved in the Urethrogenital Reflex in the Female Rat." *Journal of Comparative Neurology* 462:355–70.

Masters, W. H., and V. E. Johnson. 1966. *Human Sexual Response.* Boston: Little, Brown.

Mautz, B. S., B. B. M. Wong, R. A. Peters, and M. D. Jennions. 2013. "Penis Size Interacts with Body Shape and Height to Influence Male Attractiveness." *Proceedings of the National Academy of Sciences* 110:6925–30.

McCracken, K. G. 2000. "The 20-cm Spiny Penis of the Argentine Lake Duck (*Oxyura vittata*)." *The Auk* 117:820–25.

McCracken, K. G., R. E. Wilson, P. J. McCracken, and K. P. Johnson. 2001. "Sexual Selection: Are Ducks Impressed by Drakes' Display?" *Nature* 413:128.

McLean, C. Y., P. L. Reno, A. A. Pollen, A. I. Bassan, T. D. Capellini, C. Guenther, V. B. Indjeian, X. Lim, D. B. Menke, B. T. Schaar, A. M. Wenger, G. Bejerano, and D. M. Kingsley. 2011. "Human-Specific Loss of Regulatory DNA and the Evolution of Human-Specific Traits." *Nature* 471:216–19.

McPeek, M. A., L. Shen, and H. Farid. 2009. "The Correlated Evolution of Three-Dimensional Reproductive Structures Between Male and Female Damselflies." *Evolution* 63:73–83.

McPeek, M. A., L. Shen, J. Z. Torrey, and H. Farid. 2008. "The Tempo and Mode of Three-Dimensional Morphological Evolution in Male Reproductive Structures." *American Naturalist* 171:e158–e178.

Meisenheimer, J. 1912. "Die Weinbergschnecke *Helix pomatia* L." In *Monographien einheimischer Tiere,* edited by H. E. Ziegler and R. Wolterek, 1–140. Leipzig, Germany: Werner Klinkhardt.

Mellanby, K. 1939. "Fertilization and Egg-Production in the Bed-Bug, *Cimex lectularius* L." *Parasitology* 31:193–99.

Michalik, P., B. Knoflach, K. Thaler, and G. Alberti. 2010. "Live for the Moment: Adaptations in the Male Genital System of a Sexually Cannibalistic Spider (Theridiidae, Araneae)." *Tissue and Cell* 42:32–36.

Michiels, N. K., and J. M. Koene. 2006. "Sexual Selection Favors Harmful Mating in Hermaphrodites More Than in Gonochorists." *Integrative and Comparative Biology* 46:473–80.

Michiels, N. K., and L. J. Newman. 1998. "Sex and Violence in Hermaphrodites." *Nature* 391:647.

Milius, S. 2013. "Sea Slug Carries Disposable Penis, Plus Spares." *Science News* 183(6):9.

Moeliker, C. W. 2001. "The First Case of Homosexual Necrophilia in the Mallard *Anas platyrhynchos* (Aves: Anatidae)." *Deinsea* 8:243–47.

Moeliker, C. W. 2009. *De Eendenman.* Amsterdam: Nieuw Amsterdam.

Møller, A. P. 1990. "Effects of a Haematophagous Mite on Secondary Sexual Tail Ornaments in the Barn Swallow (*Hirundo rustica*): A Test of the Hamilton and Zuk Hypothesis." *Evolution* 44:771–84.

Morris, D. 1967. *The Naked Ape.* London: Jonathan Cape.

Morrow, E. H., and G. Arnqvist. 2003. "Costly Traumatic Insemination and a Female Counter-Adaptation in Bed Bugs." *Proceedings of the Royal Society B* 270:2377–81.

Morrow, E. H., G. Arnqvist, and S. Pitnick. 2003. "Adaptation Versus Pleiotropy: Why Do Males Harm Their Mates?" *Behavioral Ecology* 14:802–6.

Motas, C. 1966. "Hommage à la mémoire de René Jeannel." *International Journal of Speleology* 2:229–67.

Murer, V., J. F. Spetz, U. Hengst, L. M. Altrogge, A. de Agostini, and D. Monard. 2001. "Male Fertility Defects in Mice Lacking the Serine Protease Inhibitor Protease Nexin-1." *Proceedings of the National Academy of Sciences* 98:3029–33.

Myers, C. W., and J. W. Daly. 1976. "Preliminary Evaluation of Skin Toxins and Vocalization in Taxonomic and Evolutionary Studies of Poison-Dart Frogs (Dendrobatidae)." *Bulletin of the American Museum of Natural History* 157:173–262.

Naef, A. 1921. "Die Cephalopoden, Systematik." *Flora Fauna Golf. Neapel* 35:1–863.

Nahum, G. G., H. Stanislaw, and C. McMahon. 2004. "Preventing Ectopic Pregnancies: How Often Does Transperitoneal Transmigration of Sperm Occur in Effecting Human Pregnancy?" *BJOG: An International Journal of Obstetrics and Gynaecology* 111:706–14.

Negrea, Ş. 2007. "Historical Development of Biospeleology in Romania After the Death of Emile Racovitza." *Travaux de l'Institut de Spéologie «Émile Racovitza»* 45/46:131–67.

Nessler, S. H., G. Uhl, and J. M. Schneider. 2007. "Genital Damage in the Orb-Web Spider *Argiope bruennichi* (Araneae: Araneidae) Increases Paternity Success." *Behavioral Ecology* 18:174–81.

Neufeld, C. J., and A. R. Palmer. 2008. "Precisely Proportioned: Intertidal Barnacles Alter Penis Form to Suit Coastal Wave Action." *Proceedings of the Royal Society B* 275:1081–87.

Ober, C., T. Hyslop, S. Elias, L. R. Weitkamp, and W. W. Hauck. 1998. "Human Leucocyte Antigen Matching and Fetal Loss: Results of a 10-Year Prospective Study." *Human Reproduction* 13:33–38.

O'Connell, H. E., J. M. Hutson, C. R. Anderson, and R. J. Pletner. 1998. "Anatomical Relationship Between Urethra and Clitoris." *Journal of Urology* 159:1892–97.

O'Connell, H. E., K. V. Sanjeevan, and J. M. Hutson. 2005. "Anatomy of the Clitoris." *Journal of Urology* 174:1189–95.

Olsen, M. W. 1966. "Segregation and Replication of Chromosomes in Turkey Parthenogenesis." *Nature* 212:435–36.

Ono, T., M. T. Siva-Jothy, and A. Kato. 1989. "Removal and Subsequent Ingestion of Rivals' Semen During Copulation in a Tree Cricket." *Physiological Entomology* 14:195–202.

Park, G. M., J. Y. Kim, J. H. Kim, and J. K. Huh. 2012. "Penetration of the Oral Mucosa by Parasite-Like Sperm Bags of Squid: A Case Report in a Korean Woman." *Journal of Parasitology* 98:222–23.

Parker, G. A. 1970. "Sperm Competition and Its Evolutionary Consequences in the Insects." *Biological Reviews* 45:525–67.

Parker, G. A. 2001. "Golden Flies, Sunlit Meadows: A Tribute to the Yellow Dungfly." In *Model Systems in Behavioral Ecology: Integrating Conceptual, Theoretical, and Empirical Approaches,* edited by L. A. Dugatkin, 3–26. Princeton, NJ: Princeton University Press.

Partridge, L. 1988. "The Rare-Male Effect: What Is Its Evolutionary Significance?" *Philosophical Transactions of the Royal Society B* 319:525–39.

Pauls, R., G. Mutema, J. Segal, W. A. Silva, S. Kleeman, V. Dryfhout, and M. Karram. 2006. "A Prospective Study Examining the Anatomic Distribution of Nerve Density in the Human Vagina." *Journal of Sexual Medicine* 3:979–87.

Peretti, A. V., and W. G. Eberhard. 2009. "Cryptic Female Choice via Sperm Dumping Favours Male Copulatory Courtship in a Spider." *Journal of Evolutionary Biology* 23:271–81.

Perkin, A. 2007. "Comparative Penile Morphology of East African Galagos of the Genus *Galagoides* (Family Galagidae): Implications for Taxonomy." *American Journal of Primatology* 69:16–26.

Perreau, M. 2012. "Description of a New Genus and Two New Species of Leiodidae (Coleoptera) from Baltic Amber Using Phase Contrast Synchrotron X-ray Microtomography." *Zootaxa* 3455:81–88.

Perreau, M., and P. Tafforeau. 2011. "Virtual Dissection Using Phase-Contrast X-ray Synchrotron Microtomography: Reducing the Gap Between Fossils and Extant Species." *Systematic Entomology* 36:573–80.

Petrie, M. 1994. "Improved Growth and Survival of Offspring of Peacocks with More Elaborate Trains." *Nature* 371:598–99.

Pilch, B., and M. Mann. 2006. "Large-Scale and High-Confidence Proteomic Analysis of Human Seminal Plasma." *Genome Biology* 7:r40.

Pitnick, S., T. Markow, and G. S. Spicer, 1999. "Evolution of Multiple Kinds of Female Sperm Storage Organs in *Drosophila.*" *Evolution* 53:1804–22.

Place, N. J., and S. E. Glickman. 2004. "Masculinization of Female Mammals: Lessons from Nature." *Advances in Experimental Medicine and Biology* 545:243–53.

Ploog, D. W., and P. D. MacLean. 1963. "Display of Penile Erection in Squirrel Monkey (*Saimiri sciureus*)." *Animal Behaviour* 11:32–39.

Poinar, G. O. 1992. *Life in Amber.* Stanford, CA: Stanford University Press.

Polak, M., and A. Rashed. 2010. "Microscale Laser Surgery Reveals Adaptive Function of Male Intromittent Genitalia." *Proceedings of the Royal Society B* 277:1371–76.

Porto, M., A. Pissinatti, C. H. F. Burity, R. Tortelly, and L. Pissinatti. 2010. "Morphological Description of the Clitoris from the *Leontopithecus rosalia* (Linnaeus, 1766), *Leontopithecus chrysomelas* (Kuhl, 1820), and *Leontopithecus chrysopygus* (Mikan, 1823) (Primates, Platyrrhini, Callitrichidae)." *Annals of the National Academy of Medicine* 180(2):1–9.

Prasad, M. R. N. 1970. "Männliche Geschlechtsorgane." *Handbuch der Zoologie* 9(2):1–150.

Prasad Narra, H., and H. Ochman. 2006. "Of What Use Is Sex to Bacteria?" *Current Biology* 16:r705–r710.

Prum, R. O., and R. H. Torres. 2004. "Structural Colouration of Mammalian Skin: Convergent Evolution of Coherently Scattering Dermal Collagen Rays." *Journal of Experimental Biology* 207:2157–72.

Putnam, C. 1988. "A Little Knowledge Is a Wonderful Thing." *New Scientist* 120(1633):62–63.

Puts, D. A., K. Dawood, and L. L. M. Welling. 2012a. "Why Do Women Have Orgasms: An Evolutionary Analysis." *Archives of Sexual Behavior* 41:1127–43.

Puts, D. A., L. L. M. Welling, R. P. Burriss, and K. Dawood. 2012b. "Men's Masculinity and Attractiveness Predict Their Female Partners' Reported Orgasm Frequency and Timing." *Evolution and Human Behavior* 33:1–9.

Radtkey, R. R., S. C. Donnellan, R. N. Fisher, C. Moritz, K. A. Hanley, and T. J. Case. 1995. "When Species Collide: The Origin and Spread of an Asexual Species of Gecko." *Proceedings of the Royal Society B* 259:145–52.

Ramos, M., D. J. Irschick, and T. E. Christenson. 2004. "Overcoming an Evolutionary Conflict: Removal of a Reproductive Organ Greatly Increases Locomotor Performance." *Proceedings of the National Academy of Sciences* 101:4883–87.

Randerson, J., and L. Hurst. 2001. "The Uncertain Evolution of the Sexes." *Trends in Ecology and Evolution* 16:571–79.

Raverat, G. 1952. *Period Piece: A Cambridge Childhood.* London: Faber and Faber.

Redi, F. 1684. *Osservazioni intorno agli animali viventi.* Florence: Piero Matini.

Reeder, D. M. 2003. "The Potential for Cryptic Female Choice in Primates: Behavioral, Physiological, and Anatomical Considerations." In *Sexual Selection and Reproductive Competition in Primates: New Perspectives and Directions,* edited by C. B. Jones, 255–303. Norman, OK: American Society of Primatologists.

Reinhard, J., and D. M. Rowell. 2005. "Social Behaviour in an Australian Velvet Worm, *Euperipatoides rowelli* (Onychophora: Peripatopsidae)." *Journal of Zoology* 267:1–7.

Reinhardt, K., and M. T. Siva-Jothy. 2007. "Biology of the Bed Bugs (Cimicidae)." *Annual Review of Entomology* 52:351–74.

Reise, H. 2007. "A Review of Mating Behavior in Slugs of the Genus *Deroceras* (Pulmonata: Agriolimacidae)." *American Malacological Bulletin* 23:137–56.

Reise, H., and J. M. C. Hutchinson. 2002. "Penis-Biting Slugs: Wild Claims and Confusion." *Trends in Ecology and Evolution* 17:163.

Retief, T. A., N. C. Bennett, A. A. Kinahan, and P. W. Bateman. 2013. "Sexual Selection and Genital Allometry in the Hottentot Golden Mole (*Amblysomus hottentotus*)." *Mammalian Biology* 78:356–60.

Řezáč, M. 2009. "The Spider *Harpactea sadistica:* Co-Evolution of Traumatic Insemination and Complex Female Genital Morphology in Spiders." *Proceedings of the Royal Society B* 276:2697–701.

Rice, W. R. 1998. "Intergenomic Conflict, Interlocus Antagonistic Coevolution, and the Evolution of Reproductive Isolation." In *Endless Forms: Species and Speciation,* edited by D. J. Howard and S. J. Berlocher, 261–70. Oxford: Oxford University Press.

Richards, O. W. 1927. "The Specific Characters of the British Humblebees (Hymenoptera)." *Transactions of the Royal Entomological Society of London* 75:233–68.

Richmond, M. P., S. Johnson, and T. A. Markow. 2012. "Evolution of Reproductive Morphology Among Recently Diverged Taxa in the *Drosophila mojavensis* Species Cluster." *Ecology and Evolution* 2:397–408.

Ridley, M. 1993. *The Red Queen: Sex and the Evolution of Human Nature.* London: Harper.

Riemann, J. G., and B. J. Thorson. 1969. "Effect of Male Accessory Material on Oviposition and Mating by Female House Flies." *Annals of the Entomological Society of America* 62:828–34.

Rivnay, E. 1933. "The Tropisms Effecting Copulation in the Bed Bug." *Psyche* 40: 115–20.

Roach, M. 2008. *Bonk: The Curious Coupling of Science and Sex.* New York: Norton.

Roberts, E. K., A. Lu, T. J. Bergman, and J. C. Beehner. 2012. "A Bruce Effect in Wild Geladas." *Science* 335:1222–25.

Robertson, S. A., J. J. Bromfield, and K. P. Tremellen. 2003. "Seminal 'Priming' for Protection from Pre-Eclampsia: A Unifying Hypothesis." *Journal of Reproductive Immunology* 59:253–65.

Rodriguez, V., D. M. Windsor, and W. G. Eberhard. 2004. "Tortoise Beetle Genitalia and Demonstrations of a Sexually Selected Advantage for Flagellum Length in *Chelymorpha alternans* (Chrysomelidae, Cassidini, Stolaini)." In *New Developments in the Biology of Chrysomelidae,* edited by P. Jolivet, J. A. Santiago-Blay, and M. Schmitt, 739–48. The Hague: SBP Academic Publishing.

Rönn, J. L., M. Katvala, and G. Arnqvist. 2007. "Coevolution Between Harmful Male Genitalia and Female Resistance in Seed Beetles." *Proceedings of the National Academy of Sciences* 104:10921–25.

Rowlands, I. W. 1957. "Insemination of the Guinea-Pig by Intraperitoneal Injection." *Journal of Endocrinology* 16:98–106.

Rozendaal, S. 2006. "Over eitjes en dierckens." *Elsevier,* 48:86.

Rubenstein, N. M., G. R. Cunha, Y. Z. Wang, K. L. Campbell, A. J. Conley, K. C. Catania, S. E. Glickman, and N. J. Place. 2003. "Variation in Ovarian Morphology in Four Species of New World Moles with a Peniform Clitoris." *Reproduction* 126:713–19.

Sasabe, M., Y. Takami, and T. Sota. 2007. "The Genetic Basis of Interspecific Differences in Genital Morphology of Closely Related Carabid Beetles." *Heredity* 98:385–91.

Sbilordo, S. H., M. A. Schäfer, and P. I. Ward. 2009. "Sperm Release and Use at Fertilization by Yellow Dung Fly Females (*Scathophaga stercoraria*)." *Biological Journal of the Linnean Society* 98:511–18.

Schafstall, N. B. 2012. “Opportunities for Palaeoclimate Research on Coleoptera in Northwestern Europe.” M.Sc. thesis, Utrecht University, the Netherlands.

Schaller, R. 1971. “Indirect Sperm Transfer by Soil Arthropods.” *Annual Review of Entomology* 16:407–46.

Scharf, I., and O. Y. Martin. 2013. “Same-Sex Sexual Behavior in Insects and Arachnids: Prevalence, Causes, and Consequences.” *Behavioral Ecology and Sociobiology* 67:1719–30.

Schilthuizen, M. 2000. *Frogs, Flies, and Dandelions: The Making of Species.* Oxford: Oxford University Press.

Schilthuizen, M. 2001. “Slug Sex Shocker.” *Science Now* online, September 6, 2001, http://news.sciencemag.org/2001/09/slug-sex-shocker.

Schilthuizen, M. 2003. “The Race for Solid Semen.” *Science Now* online, November 24, 2003, http://news.sciencemag.org/2003/11/race-solid-semen.

Schilthuizen, M. 2004. “Why Two Sexes Are Better Than One.” *Science Now* online, October 6, 2004, http://news.sciencemag.org/2004/10/why-two-sexes-are-better-one.

Schilthuizen, M. 2005. “The Darting Game in Snails and Slugs.” *Trends in Ecology and Evolution* 20:581–84.

Schilthuizen, M. 2009. *The Loom of Life: Unravelling Ecosystems.* Berlin: Springer.

Schilthuizen, M. 2010. “Darwins Peepshow.” *Bionieuws,* 20(18):8–9.

Schilthuizen, M. 2013. “Pelgrim in Parijs.” *Entomologische Berichten* 73:41.

Schilthuizen, M., P. G. Craze, A. S. Cabanban, A. Davison, E. Gittenberger, J. Stone, and B. J. Scott. 2007. “Sexual Selection Maintains Whole-Body Chiral Dimorphism.” *Journal of Evolutionary Biology* 20:1941–49.

Schilthuizen, M., and A. Davison. 2005. “The Convoluted Evolution of Snail Chirality.” *Naturwissenschaften* 92:504–15.

Schilthuizen, M., M. Haase, K. Koops, S. Looijestijn, and S. Hendrikse. 2012. “The Ecology of Shell Shape Difference in Chirally Dimorphic Snails.” *Contributions to Zoology* 81:95–101.

Schilthuizen, M., and S. Looijestijn. 2009. “The Sexology of the Chirally Dimorphic Snail Species *Amphidromus inversus* (Gastropoda: Camaenidae).” *Malacologia* 51:379–87.

Schilthuizen, M., B. J. Scott, A. S. Cabanban, and P. G. Craze. 2005. “Population Structure and Coil Dimorphism in a Tropical Land Snail.” *Heredity* 95:216–20.

Schneider, J. M., and P. Michalik. 2011. “One-Shot Genitalia Are Not an Evolutionary Dead End: Regained Male Polygamy in a Sperm Limited Spider Species.” *BMC Evolutionary Biology* 11:e197.

Schoot, P. van der, J. van Ophemert, and R. Baumgarten. 1992. “Copulatory Stimuli in Rats Induce Heat Abbreviation Through Effects on Genitalia but Not Through Effects on Central Nervous Mechanisms Supporting the Steroid Hormone-Induced Sexual Responsiveness.” *Behavioural Brain Research* 49:213–23.

Sedgwick, A. 1885. “The Development of *Peripatus capensis.*” *Proceedings of the Royal Society* 38:354–61.

Sekizawa, A., S. Seki, M. Tokuzato, S. Shiga, and Y. Nakashima. 2013. "Disposable Penis and Its Replenishment in a Simultaneous Hermaphrodite." *Biology Letters* 9:20121150.

Shah, J., and N. Christopher. 2002. "Can Shoe Size Predict Penile Length?" *BJU International* 90:586–87.

Shapiro, A. M., and A. H. Porter. 1989. "The Lock and Key Hypothesis: Evolutionary and Biosystematic Interpretation of Insect Genitalia." *Annual Review of Entomology* 34:231–45.

Shen, L., H. Farid, and M. A. McPeek. 2009. "Modeling Three-Dimensional Morphological Structures Using Spherical Harmonics." *Evolution* 63:1003–16.

Siva-Jothy, M. T., and R. E. Hooper. 1996. "Differential Use of Stored Sperm During Oviposition in the Damselfly *Calopteryx splendens xanthostoma* (Charpentier)." *Behavioral Ecology and Sociobiology* 39:389–93.

Springer, M. S., G. C. Cleven, O. Madsen, W. W. de Jong, V. G. Waddell, H. M. Amrine, and M. J. Stanhope. 1997. "Endemic African Mammals Shake the Phylogenetic Tree." *Nature* 388:61–64.

Stockley, P. 2002. "Sperm Competition Risk and Male Genital Anatomy: Comparative Evidence for Reduced Duration of Female Sexual Receptivity in Primates with Penile Spines." *Evolutionary Ecology* 16:123–37.

Stutt, A. D., and M. T. Siva-Jothy. 2001. "Traumatic Insemination and Sexual Conflict in the Bed Bug *Cimex lectularius*." *Proceedings of the National Academy of Sciences* 98:5683–87.

Sukhsangchan, C., and J. Nabhitabhat. 2007. "Embryonic Development of Muddy Paper Nautilus, *Argonauta hians* Lightfoot, 1786, from Andaman Sea, Thailand." *Kasetsart Journal: Natural Science* 41:531–38.

Summers, K. 2004. "Cross-Breeding of Distinct Color Morphs of the Strawberry Poison Frog (*Dendrobates pumilio*) from the Bocas del Toro Archipelago, Panama." *Journal of Herpetology* 38:1–8.

Symons, D. 1979. *The Evolution of Human Sexuality.* Oxford: Oxford University Press.

Tait, N. N., and D. A. Briscoe. 1990. "Sexual Head Structures in the Onychophora: Unique Modifications for Sperm Transfer." *Journal of Natural History* 24: 1517–27.

Tait, N. N., and J. M. Norman. 2001. "Novel Mating Behaviour in *Florelliceps stutchburyae* Gen. Nov., Sp. Nov. (Onychophora: Peripatopsidae) from Australia." *Journal of Zoology* 253:301–8.

Tallamy, D. W., B. E. Powell, and J. A. McClafferty. 2001. "Male Traits Under Cryptic Female Choice in the Spotted Cucumber Beetle (Coleoptera: Chrysomelidae)." *Behavioral Ecology* 13:511–18.

Tanabe, T., and T. Sota. 2008. "Complex Copulatory Behavior and the Proximate Effect of Genital and Body Size Differences on Mechanical Reproductive Isolation in the Millipede Genus *Parafontaria*." *American Naturalist* 171:692–99.

Tatsuta, H., K. Mizota, and S.-I. Akimoto. 2001. "Allometric Patterns of Heads and Genitalia in the Stag Beetle *Lucanus maculifemoratus* (Coleoptera: Lucanidae)." *Annals of the Entomological Society of America* 94:462–66.

Tauber, P. F., L. J. D. Zaneveld, D. Propping, and G. F. B. Schumacher. 1975. "Components of Human Split Ejaculates: I. Spermatozoa, Fructose, Immunoglobulins, Albumin, Lactoferrin, Transferrin and Other Plasma Proteins." *Journal of Reproduction and Fertility* 43:249–67.

Tinbergen, L. 1939. "Zur Fortpflanzungsethologie von *Sepia officinalis* L." *Archives Néerlandaises de Zoologie* 3:323–64.

Tinklepaugh, O. L. 1930. "Occurrence of Vaginal Plug in a Chimpanzee." *Anatomical Record* 46:329–32.

Tripp, H. R. H. 1971. "Reproduction in Elephant-Shrews (Macroscelididae) with Special Reference to Ovulation and Implantation." *Journal of Reproduction and Fertility* 26:149–59.

Trivers, R. L. 1972. "Parental Investment and Sexual Selection." In *Sexual Selection and the Descent of Man 1871–1971,* edited by B. Campbell, 136–79. Chicago: Aldine.

Troisi, A., and M. Carosi. 1998. "Female Orgasm Rate Increases with Male Dominance in Japanese Macaques." *Animal Behaviour* 56:1261–66.

Tutin, C. E. G. 1979. "Mating Patterns and Reproductive Strategies in a Community of Wild Chimpanzees (*Pan troglodytes schweinfurthii*)." *Behavioral Ecology and Sociobiology* 6:29–38.

Uhl, G., S. H. Nessler, and J. M. Schneider. 2007. "Copulatory Mechanism in a Sexually Cannibalistic Spider with Genital Mutilation (Araneae: Araneidae: *Argiope bruennichi*)." *Zoology* 110:398–408.

Uhl, G., S. H. Nessler, and J. M. Schneider. 2010. "Securing Paternity in Spiders? A Review on Occurrence and Effects of Mating Plugs and Male Genital Mutilation." *Genetica* 138:75–104.

VanDemark, N. L., and R. L. Hays. 1952. "Uterine Motility Responses to Mating." *American Journal of Physiology* 170:518–21.

Veerman, E. 2010. "Onderscheidende penissen." *Noorderlicht,* September 22, 2010, www.wetenschap24.nl/nieuws/artikelen/2010/september/Onderscheidende-penissen.html.

Vetten, L., and S. Haffejee. 2005. "Gang Rape: A Study in Inner-City Johannesburg." *SA Crime Quarterly* 12:31–36.

Waage, J. K. 1979. "Dual Function of the Damselfly Penis: Sperm Removal and Transfer." *Science* 203:916–18.

Waage, J. K. 1982. "Sperm Displacement by Male *Lestes vigilax* Hagen (Odonata: Zygoptera)." *Odonatologica* 11:201–9.

Waal, F. B. M. de. 1986. "The Brutal Elimination of a Rival Among Captive Chimpanzees." *Ethology and Sociobiology* 7:237–51.

Walker, M. H., E. M. Roberts, T. Roberts, G. Spitteri, M. J. Streubig, J. L. Hartland, and N. N. Tait. 2006. "Observations on the Structure and Function of the Seminal Receptacles and Associated Accessory Pouches in Ovoviviparous Onychophorans from Australia (Peripatopsidae; Onychophora)." *Journal of Zoology* 270:531–42.

Wallen, K., and E. A. Lloyd. 2008. "Clitoral Variability Compared with Penile Variability Supports Nonadaptation of Female Orgasm." *Evolution and Development* 10:1–2.

Ward, P. I. 1993. "Females Influence Sperm Storage and Use in the Yellow Dung Fly *Scatophaga stercoraria* (L.)." *Behavioral Ecology and Sociobiology* 32:313–19.

Wasser, S. K., and D. Y. Isenberg. 1986. "Reproductive Failure Among Women: Pathology or Adaptation?" *Journal of Psychosomatic Obstetrics and Gynaecology* 5:153–75.

Watson, P. J. 1991. "Multiple Paternity as Genetic Bet-Hedging in Female Sierra Dome Spiders, *Linyphia litigiosa* (Linyphiidae)." *Animal Behaviour* 41:343–60.

Watson, P. J. 1995. "Dancing in the Dome." *Natural History* 104(3):40–43.

Watson, P. J. and J. R. B. Lighton. 1994. "Sexual Selection and the Energetics of Copulatory Courtship in the Sierra Dome Spider, *Linyphia litigiosa*." *Animal Behaviour* 48:615–26.

Weiner, J. 1994. *The Beak of the Finch: A Story of Evolution in Our Time.* New York: Knopf.

West-Eberhard, M. J. 1984. "Sexual Selection, Social Communication, and Species-Specific Signals in Insects." In *Insect Communication*, edited by T. Lewis, 283–324. London: Academic Press.

Weygoldt, P. 1969. *The Biology of Pseudoscorpions.* Cambridge, MA: Harvard University Press.

Whitney, N. M., H. L. Pratt, and J. C. Carrier. 2004. "Group Courtship, Mating Behaviour and Siphon Sac Function in the White-Tip Reef Shark, *Triaenodon obesus*." *Animal Behaviour* 68:1435–42.

Wickler, W. 1966. "Ursprung und biologische Deutung des Genitalpräsentierens männlicher Primaten." *Zeitschrift für Tierpsychologie* 23:422–37.

Wiktor, A. 1966. "Eine neue Nacktschneckenart (Gastropoda, Limacidae) aus Polen." *Annales Zoologici* 23:449–57.

Wildt, L., S. Kissler, P. Licht, and W. Becker. 1998. "Sperm Transport in the Human Female Genital Tract and Its Modulation by Oxytocin as Assessed by Hysterosalpingoscintigraphy, Hysterotonography, Electrohysterography, and Doppler Sonography." *Human Reproduction Update* 4:655–66.

Williams, P. H. 1991. "The Bumble Bees of the Kashmir Himalaya (Hymenoptera: Apidae: Bombini)." *Bulletin of the British Museum of Natural History: Entomology* 60:1–204.

Williamson, S. 1998. "The Truth About Women." *New Scientist* 159(2145):34.

Winterbottom, M., T. Burke, and T. R. Birkhead. 1999. "A Stimulatory Phalloid Organ in a Weaver Bird." *Nature* 399:28.

Winterbottom, M., T. Burke, and T. R. Birkhead. 2001. "The Phalloid Organ, Orgasm and Sperm Competition in a Polygynandrous Bird: The Red-Billed Buffalo Weaver (*Bubalornis niger*)." *Behavioral Ecology and Sociobiology* 50:474–82.

Wojcieszek, J. M., and L. W. Simmons. 2011. "Evidence for Stabilizing Selection and Slow Divergent Evolution of Male Genitalia in a Millipede (*Antichiropus variabilis*)." *Evolution* 66:1138–53.

Woodall, P. F. 1995. "The Penis of Elephant Shrews (Mammalia: Macroscelididae)." *Journal of Zoology* 237:399–410.

Yamane, T., and T. Miyatake. 2012. "Evolutionary Correlation Between Male Substances and Female Remating Frequency in a Seed Beetle." *Behavioral Ecology* 23:715–22.

Yanagimachi, R., and M. C. Chang. 1963. "Sperm Ascent Through the Oviduct of the Hamster and Rabbit in Relation to the Time of Ovulation." *Journal of Reproduction and Fertility* 6:413–20.

Yong, E. 2012. "The Bruce Effect: Why Some Pregnant Monkeys Abort When New Males Arrive." *Discover* online, February 23, 2012, http://blogs.discovermagazine.com/notrocketscience/2012/02/23/the-bruce-effect-why-some-pregnant-monkeys-abort-when-new-males-arrive/.

Zarrow, M. X., and J. H. Clark. 1968. "Ovulation Following Vaginal Stimulation in a Spontaneous Ovulator and Its Implications." *Journal of Endocrinology* 40:343–52.

Zietsch, B. P., and P. Santtila. 2011. "Genetic Analysis of Orgasmic Function in Twins and Siblings Does Not Support the By-Product Theory of Female Orgasm." *Animal Behaviour* 82:1097–1101.

Index

Page numbers in *italics* refer to illustrations.